John Wiley & Sons, Inc.

Learning Resources
Centre

12732710

To order books or for customer service call 1-800-CALL-WILEY (225-5945).

ISBN 0-471-67810-4

Printed in the United States of America

10 9 8 7 6 5 4 3 2 1

Printed and bound by Hamilton Printing

CHAPTER 1

Quality Improvement in the Modern Business Environment

Learning Objectives

After completing this chapter you should be able to:

1. Define and discuss quality and quality improvement
2. Discuss the different dimensions of quality
3. Discuss the evolution of modern quality improvement methods
4. Discuss the role that variability and statistical methods play in controlling and improving quality
5. Describe the quality management philosophies of W. Edwards Deming, Joseph M. Juran, and Armand V. Feigenbaum
6. Discuss total quality management, the Malcolm Baldrige National Quality Award, Six-Sigma, and quality systems and standards
7. Explain the links between quality and productivity and between quality and cost
8. Discuss product liability
9. Discuss the three functions: quality planning, quality assurance, and quality control and improvement

Important Terms and Concepts

Acceptance sampling
Deming's 14 points
Dimensions of quality
Internal and External failure costs
Nonconforming product or service
Product liability
Quality characteristics
Quality engineering
Quality of design
Quality systems and standards
Specifications
The Juran Trilogy
Total quality management (TQM)

Appraisal costs
Designed experiments
Fitness for use
ISO 9000:2000
Prevention costs
Quality assurance
Quality control and improvement
Quality of conformance
Quality planning
Six-Sigma
Statistical process control (SPC)
The Malcolm Baldrige National Quality Award
Variability

The modern definition of quality, "Quality is inversely proportional to variability" (text p. 4), implies that product quality increases as variability in important product characteristics decreases. Quality improvement can then be defined as "... the reduction of variability in processes and products" (text p. 6). Since the early 1900's, statistical methods have been used to control and improve quality. In the Introduction to Statistical Quality Control, 4th ed., by Douglas C. Montgomery, methods applicable in the key areas of process control, design of experiments, and acceptance sampling are presented.

To understand the potential for application of statistical methods, it may help to envision the system that creates a product as a "black box" (text Figure 1-3). The output of this black box is a product whose quality is defined by one or more quality characteristics that represent dimensions such as conformance to standards, performance, or reliability. Product quality can be evaluated with acceptance sampling plans. These plans are typically applied to either the output of a process or the input raw materials and components used in manufacturing. Application of process control techniques (such as control charts) or statistically designed experiments can achieve significant reduction in variability.

Black box inputs are categorized as "incoming raw materials and parts," "controllable inputs," and "uncontrollable inputs."

The quality of incoming raw materials and parts is often assessed with acceptance sampling plans. As material is received from suppliers, incoming lots are inspected then dispositioned as either acceptable or unacceptable. Once a history of high quality material is established, a customer may accept the supplier's process control data in lieu of incoming inspection results.

"Controllable" and "uncontrollable" inputs apply to incoming materials, process variables, and environmental factors. For example, it may be difficult to control the temperature in a heat-treating oven in the sense that some areas of the oven may be cooler while some areas may be warmer. Properties of incoming materials may be very difficult to control. For example, the moisture content and proportion of hardwood in trees used for papermaking have a significant impact on the quality characteristics of the finished paper. Environmental variables such as temperature and relative humidity are often hard to control precisely.

Whether or not controllable and uncontrollable inputs are significant can be determined through process characterization. Statistically designed experiments are extremely helpful in characterizing processes and optimizing the relationship between incoming materials, process variables, and product characteristics

Although the initial tendency is to think of manufacturing processes and products, the statistical methods presented in this text can also be applied to business processes and products, such as financial transactions and services. In some organizations the opportunity to improve quality in thee areas is even greater than it is in manufacturing.

Various quality philosophies and management systems are briefly described in the text; a common thread is the necessity for continuous improvement to increase productivity and reduce cost. The technical tools described in the text are essential for successful quality improvement. Quality management systems alone do not reduce variability and improve quality.

CHAPTER 2

Modeling Process Quality

Learning Objectives

After completing this chapter you should be able to:

1. Construct and interpret visual data displays, including the stem-and-leaf plot, the histogram, and the box plot
2. Compute and interpret the sample mean, the sample variance, the sample standard deviation, and the sample range
3. Explain the concepts of a random variable and a probability distribution
4. Understand and interpret the mean, variance, and standard deviation of a probability distribution
5. Determine probabilities from probability distributions
6. Understand the assumptions for each of the discrete probability distributions presented
7. Understand the assumptions for each of the continuous probability distributions presented
8. Select an appropriate probability distribution

Important Terms and Concepts

Approximations to probability distributions
Box plot
Central limit theorem
Discrete distribution
Gamma distribution
Histogram
Interquartile range
Mean of a distribution
Negative binomial distribution
Normal probability plot
Percentile
Population
Probability plotting
Random variable
Sample
Sample standard deviation
Standard deviation of a distribution
Statistics
Time series plot
Variance of a distribution

Binomial distribution
Continuous distribution
Descriptive statistics
Exponential distribution
Geometric distribution
Hypergeometric distribution
Lognormal distribution
Median
Normal distribution
Pascal distribution
Poisson distribution
Probability distribution
Quartile
Run chart
Sample average
Sample Variance
Standard normal distribution
Stem-and-leaf display
Uniform distribution
Weibull distribution

Exercises

2-1. The content of liquid detergent bottles is being analyzed. Twelve bottles, randomly selected from the process, are measured, and the results are as follows (in fluid ounces): 16.05, 16.03, 16.02, 16.04, 16.05, 16.01, 16.02, 16.02, 16.03, 16.01, 16.00, 16.07.

(a) Calculate the sample average.

$$\bar{x} = \sum_{i=1}^{n} x_i \Big/ n = \left(16.05 + 16.03 + \cdots + 16.07\right)\Big/12 = 16.029 \text{ oz}$$

(b) Calculate the sample standard deviation.

$$s = \sqrt{\frac{\sum_{i=1}^{n} x_i^2 - \left(\sum_{i=1}^{n} x_i\right)^2 \Big/ n}{n-1}} = \sqrt{\frac{(16.05^2 + \cdots + 16.07^2) - (16.05 + \cdots + 16.07)^2 \Big/ 12}{12-1}} = 0.0202 \text{ oz}$$

MTB > Stat > Basic Statistics > Display Descriptive Statistics

Descriptive Statistics: Ex2-1

Variable	N	N*	Mean	SE Mean	StDev	Minimum	Q1	Median	Q3
Ex2-1	12	0	16.029	0.00583	0.0202	16.000	16.013	16.025	16.048
Variable	Maximum								
Ex2-1	16.070								

2-3. The nine measurements that follow are furnace temperatures recorded on successive batches in a semiconductor manufacturing process (units are °F): 953, 955, 948, 951, 957, 949, 954, 950, 959

(a) Calculate the sample average.

$$\bar{x} = \sum_{i=1}^{n} x_i \Big/ n = \left(953 + 955 + \cdots + 959\right)\Big/9 = 952.9 \text{ °F}$$

(b) Calculate the sample standard deviation.

$$s = \sqrt{\frac{\sum_{i=1}^{n} x_i^2 - \left(\sum_{i=1}^{n} x_i\right)^2 \Big/ n}{n-1}} = \sqrt{\frac{(953^2 + \cdots + 959^2) - (953 + \cdots + 959)^2 \Big/ 9}{9-1}} = 3.7 \text{ °F}$$

MTB > Stat > Basic Statistics > Display Descriptive Statistics

Descriptive Statistics: Ex2-3

Variable	N	N*	Mean	SE Mean	StDev	Minimum	Q1	Median	Q3
Ex2-3	9	0	952.89	1.24	3.72	948.00	949.50	953.00	956.00
Variable	Maximum								
Ex2-3	959.00								

2-5. Yield strengths of circular tubes with end caps are measured. The first yields (in kN) are as follows: 96, 102, 104, 108, 126, 128, 150, 156

(a) Calculate the sample average.

$$\bar{x} = \sum_{i=1}^{n} x_i \Big/ n = (96 + 102 + \cdots + 156)/8 = 121.25 \text{ kN}$$

(b) Calculate the sample standard deviation.

$$s = \sqrt{\frac{\sum_{i=1}^{n} x_i^2 - \left(\sum_{i=1}^{n} x_i\right)^2 \Big/ n}{n-1}} = \sqrt{\frac{(96^2 + \cdots + 156^2) - (96 + \cdots + 156)^2 \Big/ 8}{8-1}} = 22.63 \text{ kN}$$

MTB > Stat > Basic Statistics > Display Descriptive Statistics

```
Descriptive Statistics: Ex2-5
Variable   N   N*    Mean   SE Mean   StDev   Minimum      Q1   Median       Q3
Ex2-5      8   0   121.25    8.00     22.63     96.00   102.50   117.00   144.50
Variable   Maximum
Ex2-5      156.00
```

2-7. The data shown here are chemical process yield readings on successive days (read down, then across). Construct a histogram for these data. Comment on the shape of the histogram. Does it resemble any of the distributions that we have discussed in this chapter?

Use $\sqrt{n} = \sqrt{90} \cong 9$ bins and **MTB > Graph > Histogram > Simple**

94.1	87.3	94.1	92.4	84.6	84.4
93.2	84.1	92.1	90.6	83.6	86.6
90.6	90.1	96.4	89.1	85.4	91.7
91.4	95.2	88.2	88.8	89.7	87.5
88.2	86.1	86.4	86.4	87.6	84.2
86.1	94.3	85.0	85.1	85.1	85.1
95.1	93.2	84.9	84.0	89.6	90.5
90.0	86.7	87.3	93.7	90.0	95.6
92.4	83.0	89.6	87.7	90.1	88.3
87.3	95.3	90.3	90.6	94.3	84.1
86.6	94.1	93.1	89.4	97.3	83.7
91.2	97.8	94.6	88.6	96.8	82.9
86.1	93.1	96.3	84.1	94.4	87.3
90.4	86.4	94.7	82.6	96.1	86.4
89.1	87.6	91.1	83.1	98.0	84.5

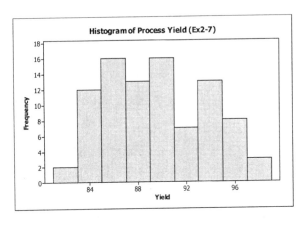

The distribution appears to have two "peaks" or central tendencies, at about 88 and 94. This is also referred to as a **bimodal** distribution. It does not resemble any of the continuous distributions discussed in this chapter (normal, lognormal, exponential, gamma, or Weibull).

2-9.   Construct and interpret a normal probability plot of the volumes of the liquid detergent bottles in Exercise 2-1.

MTB > Graph > Probability Plot > Single

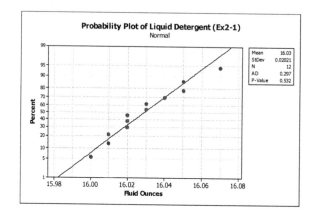

When plotted on a normal probability plot, the data points tend to fall along a straight line, indicating that a normal distribution adequately describes the volume of detergent.

2-11.   Construct a normal probability plot of the failure time data in Exercise 2-6. Does the assumption that failure time for this component is well modeled by a normal distribution seem reasonable?

MTB > Graph > Probability Plot > Single

127	124	121	118
125	123	136	131
131	120	140	125
124	119	137	133
129	128	125	141
121	133	124	125
142	137	128	140
151	124	129	131
160	142	130	129
125	123	122	126

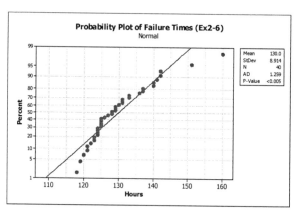

When plotted on a normal probability plot, the data points do not fall along a straight line, indicating that the normal distribution does not reasonably describe the failure times.

2-13.   Consider the viscosity data in Exercise 2-8. Construct a normal probability plot, a lognormal probability plot, and a Weibull probability plot for these data. Based on the plots, which distribution seems to be the best model for the viscosity data?

MTB > Graph > Probability Plot > Single and in the dialog box, select Distribution to choose the distributions.

13.3	14.9	15.8	16.0
14.5	13.7	13.7	14.9
15.3	15.2	15.1	13.6
15.3	14.5	13.4	15.3
14.3	15.3	14.1	14.3
14.8	15.6	14.8	15.6
15.2	15.8	14.3	16.1
14.5	13.3	14.3	13.9
14.6	14.1	16.4	15.2
14.1	15.4	16.9	14.4
14.3	15.2	14.2	14.0
16.1	15.2	16.9	14.4
13.1	15.9	14.9	13.7
15.5	16.5	15.2	13.8
12.6	14.8	14.4	15.6
14.6	15.1	15.2	14.5
14.3	17.0	14.6	12.8
15.4	14.9	16.4	16.1
15.2	14.8	14.2	16.6
16.8	14.0	15.7	15.6

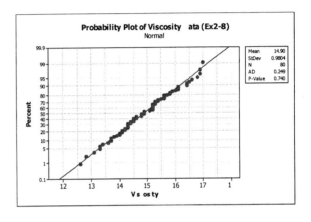

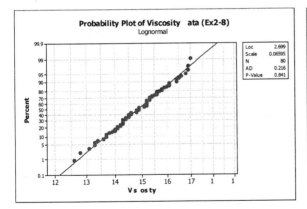

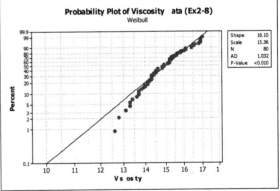

Both the normal and lognormal distributions appear to be reasonable models for the data; the plot points tend to fall along a straight line, with no bends or curves. However, the plot points on the Weibull probability plot are not straight—particularly in the tails—indicating it is not a reasonable model.

2-15.   An important quality characteristic of water is the concentration of suspended solid material (in ppm). Following are 40 measurements on suspended solids for a certain lake. Construct a normal probability plot, a lognormal probability plot, and a Weibull probability plot for these data. Based on the plots, which distribution seems to be the best model for the concentration of suspended solids?

MTB > Graph > Probability Plot > Single and in the dialog box, select Distribution to choose the distributions.

0.78	9.59	2.26	8.13	3.16
4.33	11.70	0.22	125.93	1.30
0.15	0.20	0.29	13.72	0.96
0.29	2.93	3.65	3.47	1.73
14.21	1.79	0.54	14.81	0.68
0.09	5.81	5.17	21.01	0.41
4.75	2.82	1.30	4.57	74.74
0.78	1.94	3.52	20.10	4.98

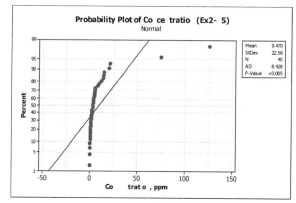

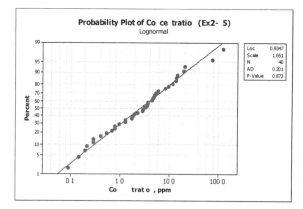

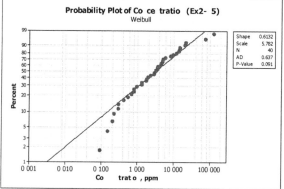

The lognormal distribution appears to be a reasonable model for the concentration data. Plotted points on the normal and Weibull probability plots tend to fall off a straight line.

2-17. Reconsider the yield data in Exercise 2-7. Construct a time-series plot for these data. Interpret the plot.

MTB > Graph > Time Series Plot > Single (or Stat > Time Series > Time Series Plot)

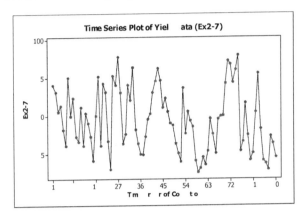

Time may be an important source of variability, as evidenced by potentially cyclic behavior.

2-19. Consider the chemical process yield data in Exercise 2-7. Calculate the sample average and standard deviation.

$$\overline{x} = \sum_{i=1}^{n} x_i \Big/ n = \frac{94.1 + 93.2 + \cdots + 84.5}{90} = 89.476$$

$$S = \sqrt{\frac{\sum_{i=1}^{n} x_i^2 - \left(\sum_{i=1}^{n} x_i\right)^2 \Big/ n}{n-1}} = \sqrt{\frac{(94.1^2 + \cdots + 84.5^2) - \frac{(94.1 + \cdots + 84.5)^2}{90}}{90-1}} = 4.158$$

MTB > Stat > Basic Statistics > Display Descriptive Statistics

Descriptive Statistics: Ex2-7

Variable	N	N*	Mean	SE Mean	StDev	Minimum	Q1	Median	Q3
Ex2-7	90	0	89.476	0.438	4.158	82.600	86.100	89.250	93.125

Variable	Maximum
Ex2-7	98.000

2-21.   Construct a box plot for the data in Exercise 2-1.

MTB > Graph > Boxplot > Simple

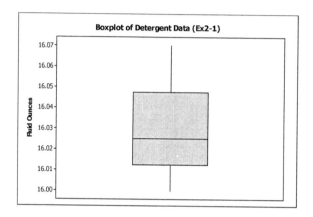

In a box plot of normally distributed data, the median line is in the middle of the box, and the two whiskers are the same length. This data is approximately normally distributed.

2-23.   Suppose that two fair dice are tossed and the random variable observed—say, x— is the sum of the two up faces. Describe the sample space of this experiment, and determine the probability distribution of x.

x: {the sum of two up faces}
sample space: {2, 3, 4, 5, 6, 7, 8, 9, 10, 11, 12}

Calculate the probability of rolling a 2: obtained by rolling a 1 on each die:
$$Pr\{x = 2\} = Pr\{1, 1\} = \frac{1}{6} \times \frac{1}{6} = \frac{1}{36}$$

Calculate the probability of rolling a 3: 1 and 2 or 2 and 1:
$$Pr\{x = 3\} = Pr\{1, 2\} + Pr\{2, 1\} = \left(\frac{1}{6} \times \frac{1}{6}\right) + \left(\frac{1}{6} \times \frac{1}{6}\right) = \frac{2}{36}$$

Calculate the probability of rolling a 4:
$$Pr\{x = 4\} = Pr\{1, 3\} + Pr\{2, 2\} + Pr\{3, 1\} = \left(\frac{1}{6} \times \frac{1}{6}\right) + \left(\frac{1}{6} \times \frac{1}{6}\right) + \left(\frac{1}{6} \times \frac{1}{6}\right) = \frac{3}{36}$$

. . .

$$p(x) = \begin{cases} 1/36; x = 2 & 2/36; x = 3 & 3/36; x = 4 & 4/36; x = 5 & 5/36; x = 6 & 6/36; x = 7 \\ 5/36; x = 8 & 4/36; x = 9 & 3/36; x = 10 & 2/36; x = 11 & 1/36; x = 12 & 0; \text{ otherwise} \end{cases}$$

2-25. A mechatronic assembly is subjected to a final functional test. Suppose that defects occur at random in these assemblies, and that defects occur according to a Poisson distribution with parameter $\lambda = 0.02$.

(a) What is the probability that an assembly will have exactly one defect?
This is a Poisson distribution with parameter $\lambda = 0.02$, $x \sim \text{POI}(0.02)$.

$$\Pr\{x = 1\} = p(1) = \frac{e^{-0.02}(0.02)^1}{1!} = 0.0196$$

(b) What is the probability that an assembly will have one or more defects?

$$\Pr\{x \geq 1\} = 1 - \Pr\{x = 0\} = 1 - p(0) = 1 - \frac{e^{-0.02}(0.02)^0}{0!} = 1 - 0.9802 = 0.0198$$

(c) Suppose that you improve the process so that the occurrence rate of defects is cut in half to $\lambda = 0.01$. What effect does this have on the probability that an assembly will have one or more defects?
This is a Poisson distribution with parameter $\lambda = 0.01$, $x \sim \text{POI}(0.01)$.

$$\Pr\{x \geq 1\} = 1 - \Pr\{x = 0\} = 1 - p(0) = 1 - \frac{e^{-0.01}(0.01)^0}{0!} = 1 - 0.9900 = 0.0100$$

Cutting the rate at which defects occur reduces the probability of one or more defects by approximately one-half, from 0.0198 to 0.0100.

2-27. The random variable x takes on the values 1, 2, or 3 with probabilities $(1 + 3k)/3$, $(1 + 2k)/3$, and $(0.5 + 5k)/3$, respectively.

$$p(x) = \{(1+3k)/3; \ x = 1 \quad (1+2k)/3; \ x = 2 \quad (0.5+5k)/3; \ x = 3 \quad 0; \ \text{otherwise}$$

(a) Find the appropriate value of k.

To solve for k, use $F(x) = \sum_{i=1}^{\infty} p(x_i) = 1$

$$\frac{(1+3k)+(1+2k)+(0.5+5k)}{3} = 1$$

$$10k = 0.5$$

$$k = 0.05$$

(b) Find the mean and variance of x.

$$\mu = \sum_{i=1}^{3} x_i p(x_i) = 1 \times \left[\frac{1+3(0.05)}{3}\right] + 2 \times \left[\frac{1+2(0.05)}{3}\right] + 3 \times \left[\frac{0.5+5(0.05)}{3}\right] = 1.867$$

$$\sigma^2 = \sum_{i=1}^{3} x_i^2 p(x_i) - \mu^2 = 1^2(0.383) + 2^2(0.367) + 3^2(0.250) - 1.867^2 = 0.615$$

(c) Find the cumulative distribution function.

$$F(x) = \begin{cases} \frac{1.15}{3} = 0.383; x = 1 \quad \frac{1.15+1.1}{3} = 0.750; x = 2 \quad \frac{1.15+1.1+0.75}{3} = 1.000; x = 3 \end{cases}$$

2-29.   A manufacturer of electronic calculators offers a one-year warranty. If the calculator fails for any reason during this period, it is replaced. The time to failure is well modeled by the following probability distribution: $f(x) = 0.125 \, e{-0.125^x} ; \quad x > 0$

(a) What percentage of the calculators will fail within the warranty period?

This is an exponential distribution with parameter $\lambda = 0.125$: $\Pr\{x \leq 1\} = F(1) = 1 - e^{-0.125(1)} = 0.118$

Approximately 11.8% will fail during the first year.

(b) The manufacturing cost of a calculator is $50, and the profit per sale is $25. What is the effect of warranty replacement on profit?

Mfg. cost = $50/calculator

Sale profit = $25/calculator

Net profit = $[-50(1 + 0.118) + 75]/calculator = $19.10/calculator.

The effect of warranty replacements is to decrease profit by $5.90/calculator.

2-31.   A production process operates with 1% nonconforming output. Every hour a sample of 25 units of product is taken, and the number of nonconforming units counted. If one or more nonconforming units are found, the process is stopped and the quality control technician must search for the cause of nonconforming production. Evaluate the performance of this decision rule.

This is a binomial distribution with parameter $p = 0.01$ and $n = 25$. The process is stopped if $x \geq 1$.

$$\Pr\{x \geq 1\} = 1 - \Pr\{x < 1\} = 1 - \Pr\{x = 0\} = 1 - \binom{25}{0}(0.01)^0(1 - 0.01)^{25} = 1 - 0.78 = 0.22$$

This decision rule means that 22% of the samples will have one or more nonconforming units, and the process will be stopped to look for a cause. This is a somewhat difficult operating situation.

This exercise may also be solved using Excel (Excel Function BINOMDIST(x, n, p, TRUE)) or MINITAB.

MTB > Calc > Probability Distributions > Binomial
Cumulative Distribution Function
```
Binomial with n = 25 and p = 0.01
x    P( X <= x )
0       0.777821
```

2-33.   A random sample of 50 units is drawn from a production process every half hour. The fraction of nonconforming product manufactured is 0.02. What is the probability that $\hat{p} \leq 0.04$ if the fraction nonconforming really is 0.02?

This is a binomial distribution with parameter $p = 0.02$ and $n = 50$.

$$\Pr\{\hat{p} \leq 0.04\} = \Pr\{x \leq 2\} = \sum_{x=0}^{4} \binom{50}{x}(0.02)^x(1 - 0.02)^{(50-x)}$$

$$= \binom{50}{0}(0.02)^0(1 - 0.02)^{50} + \binom{50}{1}(0.02)^1(1 - 0.02)^{49} + \cdots + \binom{50}{4}(0.02)^4(1 - 0.02)^{46} = 0.921$$

Therefore, the probability is 0.921 that the sample fraction nonconforming can be $\hat{p} \leq 0.04$ if the population fraction nonconforming really is $p = 0.02$.

2-35.   An electronic component for a medical x-ray unit is produced in lots of size $N = 25$. An acceptance testing procedure is used by the purchaser to protect against lots that contain too many nonconforming components. The procedure consists of selecting five components at random from the lot (without replacement) and testing them. If none of the components is nonconforming, the lot is accepted.

(a) If the lot contains two nonconforming components, what is the probability of lot acceptance?

This is a hypergeometric distribution with $N = 25$ and $n = 5$, without replacement.

Given $D = 2$ and $x = 0$:

$$\Pr\{\text{Acceptance}\} = p(0) = \frac{\binom{2}{0}\binom{25-2}{5-0}}{\binom{25}{5}} = \frac{(1)(33,649)}{(53,130)} = 0.633$$

This exercise may also be solved using Excel (Excel Function HYPGEOMDIST(x, n, D, N)) or MINITAB.

MTB > Calc > Probability Distributions > Hypergeometric

Cumulative Distribution Function
```
Hypergeometric with N = 25, M = 2, and n = 5
x    P( X <= x )
0       0.633333
```

(b) Calculate the desired probability in (a) using the binomial approximation. Is this approximation satisfactory? Why or why not?

For the binomial approximation to the hypergeometric, $p = D/N = 2/25 = 0.08$ and $n = 5$.

$$\Pr\{\text{acceptance}\} = p(0) = \binom{5}{0}(0.08)^0(1-0.08)^5 = 0.659$$

This approximation, though close to the exact solution for $x = 0$, violates the rule-of-thumb that $n/N = 5/25 = 0.20$ be less than the suggested 0.1. The binomial approximation is not satisfactory in this case.

(c) Suppose the lot size was $N = 150$. Would the binomial approximation be satisfactory in this case?

For $N = 150$, $n/N = 5/150 = 0.033 \le 0.1$, so the binomial approximation would be a satisfactory approximation the hypergeometric in this case.

(d) Suppose that the purchaser will reject the lot with the decision rule of finding one or more nonconforming components in a sample of size n, and wants the lot to be rejected with probability at least 0.95 if the lot contains five or more nonconforming components. How large should the sample size n be?

Find n to satisfy $\Pr\{x \ge 1 \mid D \ge 5\} \ge 0.95$, or equivalently $\Pr\{x = 0 \mid D = 5\} < 0.05$.

$$p(0) = \frac{\binom{5}{0}\binom{25-5}{n-0}}{\binom{25}{n}} = \frac{\binom{5}{0}\binom{20}{n}}{\binom{25}{n}}$$

2-35 (d) continued

try $n = 10$

$$p(0) = \frac{\binom{5}{0}\binom{20}{10}}{\binom{25}{10}} = \frac{(1)(184,756)}{(3,268,760)} = 0.057$$

try $n = 11$

$$p(0) = \frac{\binom{5}{0}\binom{20}{11}}{\binom{25}{11}} = \frac{(1)(167,960)}{(4,457,400)} = 0.038$$

Let sample size $n = 11$.

2-37.   A textbook has 500 pages on which typographical errors could occur. Suppose that there are exactly 10 such errors randomly located on those pages. Find the probability that a random selection of 50 pages will contain no errors. Find the probability that 50 randomly selected pages will contain at least two errors.

This is a hypergeometric distribution with $N = 500$ pages, $n = 50$ pages, and $D = 10$ errors. Checking $n/N = 50/500 = 0.1 \leq 0.1$, the binomial distribution can be used to approximate the hypergeometric, with $p = D/N = 10/500 = 0.020$.

$$\Pr\{x = 0\} = p(0) = \binom{50}{0}(0.020)^0(1 - 0.020)^{50-0} = (1)(1)(0.364) = 0.364$$

$$\Pr\{x \geq 2\} = 1 - \Pr\{x \leq 1\} = 1 - [\Pr\{x = 0\} + \Pr\{x = 1\}] = 1 - p(0) - p(1)$$

$$= 1 - 0.364 - \binom{50}{1}(0.020)^1(1 - 0.020)^{50-1} = 1 - 0.364 - 0.372 = 0.264$$

2-39.   Glass bottles are formed by pouring molten glass into a mold. The molten glass is prepared in a furnace lined with firebrick. As the firebrick wears, small pieces of brick are mixed into the molten glass and finally appear as defects (called "stones") in the bottle. If we can assume that stones occur randomly at the rate of 0.00001 per bottle, what is the probability that a bottle selected at random will contain at least one such defect?

This is a Poisson distribution with $\lambda = 0.00001$ stones/bottle.

$$\Pr\{x \geq 1\} = 1 - \Pr\{x = 0\} = 1 - \frac{e^{-0.00001}(0.00001)^0}{0!} = 1 - 0.99999 = 0.00001$$

2-41.   A production process operates in one of two states: the in-control state, in which most of the units produced conform to specifications, and an out-of-control state, in which most of the units produced are defective. The process will shift from the in-control to the out-of-control state at random. Every hour, a quality control technician checks the process, and if it is in the out-of-control state, the technician detects this with probability p. Assume that when the process shifts out of control it does so immediately following a check by the inspector, and once a shift has occurred, the process cannot automatically correct itself. If t denotes the number of periods the process remains out of control following a shift before detection, find the probability distribution of t. Find the mean number of periods the process will remain in the out-of-control state.

The distribution for t is based on the number of periods for which the process shifted from in control to out of control, 1, with probability $(p)^1$, and the periods from which the process remains in the out of control state without a shift, $t - 1$, with probability $(1 - p)^{t-1}$. There is only one permutation for this situation. Hence, the distribution is the combination of the two states as follows: $\Pr(t) = p(1 - p)^{t-1}$; $t = 1, 2, 3, \ldots$
The mean is calculated using (2-5b):

$$\mu = \sum_{i=1}^{\infty} x_i p(x_i) \Rightarrow \sum_{t=1}^{\infty} t\left[p(1-p)^{t-1}\right] = p\frac{d}{dq}\left[\sum_{t=1}^{\infty} q^t\right] = \frac{1}{p}$$

2-43.   The tensile strength of a metal part is normally distributed with mean 40 lb and standard deviation 5 lb. If 50,000 parts are produced, how many would you expect to fail to meet a minimum specification limit of 35 lb tensile strength? How many would have a tensile strength in excess of 48 lb?

$x \sim N(40, 5^2)$; $n = 50,000$

How many fail the minimum specification, LSL = 35 lb.? Using the standard normal distribution, Appendix II:

$$\Pr\{x \le 35\} = \Pr\left\{z \le \frac{35 - 40}{5}\right\} = \Pr\{z \le -1\} = \Phi(-1) = 0.159$$

The number that fail the minimum specification are $(50,000) \times (0.159) = 7950$.

This exercise may also be solved using Excel (Excel Function NORMDIST(X, μ, σ, TRUE)) or MINITAB.
MTB > Calc > Probability Distributions > Normal

Cumulative Distribution Function
```
Normal with mean = 40 and standard deviation = 5
  x   P( X <= x )
 35     0.158655
```

How many exceed 48 lb.?

$$\Pr\{x > 48\} = 1 - \Pr\{x \le 48\} = 1 - \Pr\left\{z \le \frac{48 - 40}{5}\right\} = 1 - \Pr\{z \le 1.6\} = 1 - \Phi(1.6) = 1 - 0.945 = 0.055$$

The number that exceed 48 lb are $(50,000) \times (0.055) = 2750$.

2-45.   **Continuation of Exercise 2-44.** Reconsider the power supply manufacturing process in Exercise 2-44. Suppose we wanted to improve the process. Can shifting the mean reduce the number of nonconforming units produced? How much would the process variability need to be reduced in order to have all but one out of 1000 units conform to the specifications?

> **Exercise 2-44.**   The output voltage of a power supply is normally distributed with mean 5 V and standard deviation 0.02 V. If the lower and upper specifications for voltage are 4.95 V and 5.05 V, respectively, what is the probability that a power supply selected at random will conform to the specifications on voltage?

The process, with mean 5 V, is currently centered between the specification limits (target = 5 V). Shifting the process mean in either direction would increase the number of nonconformities produced. Desire $\Pr\{\text{Conformance}\} = 1 / 1000 = 0.001$. Assume that the process remains centered between the specification limits at 5 V. Need $\Pr\{x \le \text{LSL}\} = 0.001 / 2 = 0.0005$.

$$\Phi(z) = 0.0005$$

$$z = \Phi^{-1}(0.0005) = -3.29$$

$$z = \frac{\text{LSL} - \mu}{\sigma}, \quad \text{so } \sigma = \frac{\text{LSL} - \mu}{z} = \frac{4.95 - 5}{-3.29} = 0.015$$

Process variance must be reduced to 0.015^2 to have at least 999 of 1000 conform to specification.

2-47.   The life of an automotive battery is normally distributed with mean 900 days and standard deviation 35 days. What fraction of these batteries would be expected to survive beyond 1000 days?

$$x \sim N(900, 35^2)$$
$$\Pr\{x > 1000\} = 1 - \Pr\{x \le 1000\}$$

$$= 1 - \Pr\left\{x \le \frac{1000 - 900}{35}\right\}$$

$$= 1 - \Phi(2.8571)$$

$$= 1 - 0.9979$$

$$= 0.0021$$

The percentage expected to survive more than 1000 days is 0.21%.

2-49. The specifications on an electronic component in a target-acquisition system are that its life must be between 5000 and 10,000 h. The life is normally distributed with mean 7500 h. The manufacturer realizes a price of $10 per unit produced; however, defective units must be replaced at a cost of $5 to the manufacturer. Two different manufacturing processes can be used, both of which have the same mean life. However, the standard deviation of life for process 1 is 1000 h, whereas for process 2 it is only 500 h. Production costs for process 2 are twice those for process 1. What value of production costs will determine the selection between processes 1 and 2?

$x_1 \sim N(7500, \sigma_1^2 = 1000^2)$; $x_2 \sim N(7500, \sigma_2^2 = 500^2)$; LSL = 5,000 h; USL = 10,000 h
sales = $10/unit, defect = $5/unit, profit = $10 \times \Pr\{good\} + \$5 \times \Pr\{bad\} - c$

<u>For Process 1</u>
proportion defective = $p_1 = 1 - \Pr\{LSL \le x_1 \le USL\} = 1 - \Pr\{x_1 \le USL\} + \Pr\{x_1 \le LSL\}$

$$= 1 - \Pr\left\{z_1 \le \frac{10,000 - 7,500}{1,000}\right\} + \Pr\left\{z_1 \le \frac{5,000 - 7,500}{1,000}\right\}$$

$$= 1 - \Phi(2.5) + \Phi(-2.5) = 1 - 0.9938 + 0.0062 = 0.0124$$

profit for process 1 = $10 (1 - 0.0124) + 5 (0.0124) - c_1 = 9.9380 - c_1$

<u>For Process 2</u>
proportion defective = $p_2 = 1 - \Pr\{LSL \le x_2 \le USL\} = 1 - \Pr\{x_2 \le USL\} + \Pr\{x_2 \le LSL\}$

$$= 1 - \Pr\left\{z_2 \le \frac{10,000 - 7,500}{500}\right\} + \Pr\left\{z_2 \le \frac{5,000 - 7,500}{500}\right\}$$

$$= 1 - \Phi(5) + \Phi(-5) = 1 - 1.0000 + 0.0000 = 0.0000$$

profit for process 2 = $10 (1 - 0.0000) + 5 (0.0000) - c_2 = 10 - c_2$

If $c_2 > c_1 + 0.0620$, then choose process 1

CHAPTER 3

Inferences about Process Quality

Learning Objectives

After completing this chapter you should be able to:
1. Explain the concept of random sampling
2. Explain the concept of a sampling distribution
3. Explain the general concept of estimating the parameters of a population or probability distribution
4. Know how to explain the precision with which a parameter is estimated
5. Construct and interpret confidence intervals on a single mean and on the difference in two means
6. Construct and interpret confidence intervals on a single variance or the ratio of two variances
7. Construct and interpret confidence intervals on a single proportion and on the difference in two proportions
8. Test hypotheses on a single mean and on the differences in two means
9. Test hypotheses on a single variance and on the ratio of two variances
10. Test hypotheses on a single proportion and on the difference in two proportions
11. Use the P-value approach for hypothesis testing
12. Understand how the analysis of variance (ANOVA) is used to test hypotheses about the equality of more than two means

Important Terms and Concepts

Alternative hypothesis
Binomial distribution
Chi-square distribution
Confidence intervals on mean(s), known variance(s)
Confidence intervals on proportions
Confidence intervals on the variances of two normal distributions
F-distribution
Minimum variance estimator
Parameters of a distribution
Poisson distribution
Random sample
Statistic
Test statistic
Tests of hypotheses on mean(s), unknown variance(s)
Tests of hypotheses on the variance of a normal distribution
Type I error
Unbiased estimator

Analysis of variance (ANOVA)
Checking assumptions for statistical inference procedures
Confidence interval
Confidence intervals on mean(s), unknown variance(s)
Confidence intervals on the variance of a normal distribution
Critical region for a test statistic

Hypothesis testing
Null hypothesis
Point estimator
Power of a statistical test
Sampling distribution
t-distribution
Tests of hypotheses on mean(s), known variance(s)
Tests of hypotheses on proportions
Tests of hypotheses on the variances of two normal distributions
Type II error

Exercises

3-1.   The inside diameters of bearings used in an aircraft landing gear assembly are known to have a standard deviation of $\sigma = 0.002$ cm. A random sample of 15 bearings has an average inside diameter of 8.2535 cm.

(a) Test the hypothesis that the mean inside bearing diameter is 8.25 cm. Use a two-sided alternative and $\alpha = 0.05$. Since σ is known, use the standard normal distribution:

$x \sim N(\mu, \sigma)$, $n = 15$, $\overline{x} = 8.2535$ cm, $\sigma = 0.002$ cm, $\mu_0 = 8.25$, $\alpha = 0.05$

Test $H_0: \mu = 8.25$ vs. $H_1: \mu \neq 8.25$. Reject H_0 if $|Z_0| > Z_{\alpha/2}$.

$$Z_0 = \frac{\overline{x} - \mu_0}{\sigma/\sqrt{n}} = \frac{8.2535 - 8.25}{0.002/\sqrt{15}} = 6.78 \quad \text{(Equation 3-23)}$$

$Z_{\alpha/2} = Z_{0.05/2} = Z_{0.025} = 1.96$ (from Appendix II)

Reject $H_0: \mu = 8.25$, and conclude that the mean bearing ID is not equal to 8.25 cm.

(b) Find the P-value for this test.

From Appendix II, find the corresponding cumulative standard normal, $\Phi(Z_0)$ for $Z0 = 6.78$, and calculate:

P-value $= 2[1 - \Phi(Z_0)] = 2[1 - \Phi(6.78)] = 2[1 - 1.00000] = 0$

(c) Construct a 95% two-sided confidence interval on mean bearing diameter.

$$\overline{x} - Z_{\alpha/2}\left(\sigma/\sqrt{n}\right) \leq \mu \leq \overline{x} + Z_{\alpha/2}\left(\sigma/\sqrt{n}\right)$$

$$8.25 - 1.96\left(0.002/\sqrt{15}\right) \leq \mu \leq 8.25 + 1.96\left(0.002/\sqrt{15}\right) \quad \text{(Equation 3-29)}$$

$$8.249 \leq \mu \leq 8.251$$

MTB>Stat>Basic Statistics>1-Sample Z>Summarized data

```
One-Sample Z
Test of mu = 8.2535 vs not = 8.2535
The assumed standard deviation = 0.002
 N     Mean   SE Mean        95% CI            Z       P
15   8.25000  0.00052  (8.24899, 8.25101)  -6.78   0.000
```

3-3.   The service life of a battery used in a cardiac pacemaker is assumed to be normally distributed. A random sample of 10 batteries is subjected to an accelerated life test by running them continuously at an elevated temperature until failure, and the following lifetimes (in hours) are obtained: 25.5, 26.1, 26.8, 23.2, 24.2, 28.4, 25.0, 27.8, 27.3, and 25.7.

(a) The manufacturer wants to be certain that the mean battery life exceeds 25 h. What conclusions can be drawn from these data (use $\alpha = 0.05$)?

Since σ is unknown, calculate the sample standard deviation S, and use the t-distribution, one-sided:

$x \sim N(\mu, \sigma)$, $n = 10$, $\overline{x} = 26.0$, $s = 1.62$, $\mu_0 = 25$, $\alpha = 0.05$

Test $H_0: \mu = 25$ vs. $H_1: \mu > 25$. Reject H_0 if $t_0 > t_\alpha$.

$$t_0 = (\overline{x} - \mu_0)/(S/\sqrt{n}) = (26.0 - 25)/(1.62/\sqrt{10}) = 1.952 \quad \text{(Equation 3-33)}$$

$t_{\alpha, n-1} = t_{0.05, 10-1} = 1.833$ (from Appendix IV)

Reject $H_0: \mu = 25$, and conclude that the mean life exceeds 25 h.

MTB>Stat>Basic Statistics>1-Sample t>Samples in columns

```
One-Sample T: Ex3-3
Test of mu = 25 vs>25
                                          95%
                                        Lower
Variable    N     Mean    StDev  SE Mean   Bound     T      P
Ex3-3      10   26.0000   1.6248  0.5138  25.0581  1.95  0.042
```

(b) Construct a 90% two-sided confidence interval on mean life in the accelerated test.

$$\overline{x} - t_{\alpha/2, n-1}\, S/\sqrt{n} \le \mu \le \overline{x} + t_{\alpha/2, n-1}\, S/\sqrt{n}$$

$$26.0 - 1.833\left(1.62/\sqrt{10}\right) \le \mu \le 26.0 + 1.833\left(1.62/\sqrt{10}\right) \quad \text{(Equation 3-34)}$$

$$25.06 \le \mu \le 26.94$$

MTB>Stat>Basic Statistics>1-Sample t>Samples in columns

```
One-Sample T: Ex3-3
Test of mu = 25 vs not = 25
Variable    N     Mean    StDev  SE Mean        90% CI           T      P
Ex3-3      10   26.0000   1.6248  0.5138  (25.0581, 26.9419)  1.95  0.083
```

(c) Construct a normal probability plot of the battery life data. What conclusions can you draw?

MTB>Graph>Probability Plot>Single

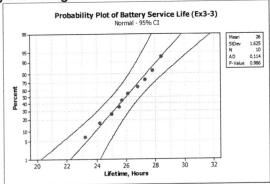

The plotted points fall approximately along a straight line, so the assumption that battery life is normally distributed is appropriate.

3-5. A new process has been developed for applying photoresist to 125-mm silicon wafers used in manufacturing integrated circuits. Ten wafers were tested, and the following photoresist thickness measurements (in angstroms × 1000) were observed: 13.3987, 13.3957, 13.3902, 13.4015, 13.4001, 13.3918, 13.3965, and 13.3925.

(a) Test the hypothesis that mean thickness is 13.4×1000 Å. Use $\alpha = 0.05$ and assume a two-sided alternative. Since σ is unknown, calculate the sample standard deviation S, and use the t-distribution:

$x \sim N(\mu, \sigma)$, $n = 10$, $\bar{x} = 13.39618 \times 1000$ Å, $s = 0.00391$, $\mu_0 = 13.4 \times 1000$ Å, $\alpha = 0.05$

Test $H_0: \mu = 13.4$ vs. $H_1: \mu \neq 13.4$. Reject H_0 if $|t_0| > t_{\alpha/2}$.

$$t_0 = (\bar{x} - \mu_0)/(S/\sqrt{n}) = (13.39618 - 13.4)/(0.00391/\sqrt{10}) = -3.089 \quad \text{(Equation 3-33)}$$

$t_{\alpha/2, n-1} = t_{0.025, 9} = 2.262$ (from Appendix IV)

Reject $H_0: \mu = 13.4$, and conclude that the mean thickness differs from 13.4×1000 Å.

MTB>Stat>Basic Statistics>1-Sample t>Samples in columns

One-Sample T: Ex3-5

Test of mu = 13.4 vs not = 13.4

Variable	N	Mean	StDev	SE Mean	95% CI	T	P
Ex3-5	10	13.3962	0.0039	0.0012	(13.3934, 13.3990)	-3.09	0.013

(b) Find a 99% two-sided confidence interval on mean photoresist thickness. Assume that thickness is normally distributed.

$$\bar{x} - t_{\alpha/2, n-1}\left(S/\sqrt{n}\right) \leq \mu \leq \bar{x} + t_{\alpha/2, n-1}\left(S/\sqrt{n}\right)$$

$$13.39618 - 3.2498\left(0.00391/\sqrt{10}\right) \leq \mu \leq 13.39618 + 3.2498\left(0.00391/\sqrt{10}\right) \quad \text{(Equation 3-34)}$$

$$13.39216 \leq \mu \leq 13.40020$$

MTB>Stat>Basic Statistics>1-Sample t>Samples in columns

One-Sample T: Ex3-5

Test of mu = 13.4 vs not = 13.4

Variable	N	Mean	StDev	SE Mean	99% CI	T	P
Ex3-5	10	13.3962	0.0039	0.0012	(13.3922, 13.4002)	-3.09	0.013

(c) Does the normality assumption seem reasonable for these data?

MTB>Graph>Probability Plot>Single

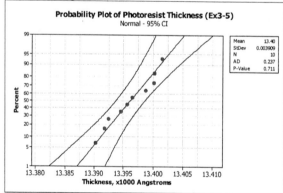

The plotted points form a reverse-"S" shape, instead of a straight line, so the assumption that battery life is normally distributed is not appropriate.

3-7. Ferric chloride is used as a flux in some types of extraction metallurgy processes. This material is shipped in containers, and the container weight varies. It is important to obtain an accurate estimate of mean container weight. Suppose that from long experience a reliable value for the standard deviation of flux container weight is determined to be 4 lb. How large a sample would be required to construct a 95% two-sided confidence interval on the mean that has a total width of 1 lb?

$\sigma = 4$ lb, $\alpha = 0.05$, $Z_{\alpha/2} = Z_{0.025} = 1.9600$, total confidence interval width = 1 lb, find n

$$2\left[Z_{\alpha/2}\left(\sigma/\sqrt{n}\right)\right] = \text{total width}$$

$$2\left[1.9600\left(4/\sqrt{n}\right)\right] = 1$$

$$n = 246$$

3-9. The output voltage of a power supply is assumed to be normally distributed. Sixteen observations taken at random on voltage are as follows: 10.35, 9.30, 10.00, 9.96, 11.65, 12.00, 11.25, 9.58, 11.54, 9.95, 10.28, 8.37, 10.44, 9.25, 9.38, and 10.85.

(a) Test the hypothesis that the mean output voltage equals 12 V against a two-sided alternative using $\alpha = 0.05$.

$x \sim N(\mu, \sigma)$, $n = 16$, $\bar{x} = 10.259$ V, $s = 0.999$ V, $\mu_0 = 12$, $\alpha = 0.05$

Test $H_0: \mu = 12$ vs. $H_1: \mu \neq 12$. Reject H_0 if $|t_0| > t_{\alpha/2}$.

$$t_0 = \left(\bar{x} - \mu_0\right)/\left(s/\sqrt{n}\right) = (10.259 - 12)/\left(0.999/\sqrt{16}\right) = -6.971 \quad \text{(Equation 3-33)}$$

$t_{\alpha/2, n-1} = t_{0.025, 15} = 2.131$ (from Appendix IV)

Reject $H_0: \mu = 12$, and conclude that the mean output voltage differs from 12V.

MTB>Stat>Basic Statistics>1-Sample t>Samples in columns

```
One-Sample T: Ex3-9
Test of mu = 12 vs not = 12
Variable    N    Mean    StDev   SE Mean      95% CI           T      P
Ex3-9      16   10.2594  0.9990   0.2498   (9.7270, 10.7917)  -6.97  0.000
```

(b) Construct a 95% two-sided confidence interval on μ.

$$\bar{x} - t_{\alpha/2, n-1}\left(s/\sqrt{n}\right) \leq \mu \leq \bar{x} + t_{\alpha/2, n-1}\left(s/\sqrt{n}\right)$$

$$10.259 - 2.131\left(0.999/\sqrt{16}\right) \leq \mu \leq 10.259 + 2.131\left(0.999/\sqrt{16}\right) \quad \text{(Equation 3-34)}$$

$$9.727 \leq \mu \leq 10.792$$

(c) Test the hypothesis that $\sigma^2 = 1$ using $\alpha = 0.05$.

$\sigma_0^2 = 1$, $\alpha = 0.05$

Test $H_0: \sigma^2 = 1$ vs. $H_1: \sigma^2 \neq 1$. Reject H_0 if $\chi^2_0 > \chi^2_{\alpha/2, n-1}$ or $\chi^2_0 < \chi^2_{1-\alpha/2, n-1}$.

$$\chi^2_0 = (n-1)S^2/\sigma_0^2 = \frac{(16-1)0.999^2}{1} = 14.970 \quad \text{(Equation 3-38)}$$

$\chi^2_{\alpha/2, n-1} = \chi^2_{0.025, 16-1} = 27.488$ (from Appendix III)

$\chi^2_{1-\alpha/2, n-1} = \chi^2_{0.975, 16-1} = 6.262$ (from Appendix III)

Do not reject $H_0: \sigma^2 = 1$, and conclude that there is insufficient evidence that the variance differs from 1.

3-9 continued

(d) Construct a 95% two-sided confidence interval on σ.

$$(n-1)S^2 \Big/ \chi^2_{\alpha/2,n-1} \leq \sigma^2 \leq (n-1)S^2 \Big/ \chi^2_{1-\alpha/2,n-1}$$

$$(16-1)0.999^2 \Big/ 27.488 \leq \sigma^2 \leq (16-1)0.999^2 \Big/ 6.262 \quad \text{(Equation 3-39)}$$

$$0.545 \leq \sigma^2 \leq 2.391$$

$$0.738 \leq \sigma \leq 1.546$$

Since the 95% confidence interval on σ contains the hypothesized value, $\sigma_0^2 = 1$, the null hypothesis, H_0: $\sigma^2 = 1$, cannot be rejected. Using MINITAB:

MTB>Stat>Basic Statistics>Graphical Summary

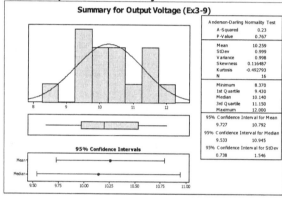

(e) Construct a 95% upper confidence interval on σ.

$$\alpha = 0.05; \quad \chi^2_{1-\alpha,n-1} = \chi^2_{0.95,15} = 7.2609 \quad \text{(from Appendix III)}$$

$$\sigma^2 \leq (n-1)S^2 \Big/ \chi^2_{1-\alpha,n-1}$$

$$\sigma^2 \leq (16-1)0.999^2 \Big/ 7.2609 \quad \text{(Equation 3-40)}$$

$$\sigma^2 \leq 2.062$$

$$\sigma \leq 1.436$$

(f) Does the assumption of normality seem reasonable for the output voltage?

MTB>Graph>Probability Plot>Single

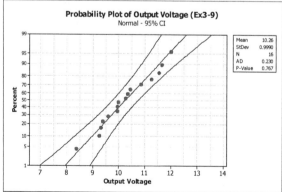

Looking at the plot, the assumption of a normal distribution for output voltage seems appropriate.

3-11.   Two quality control technicians measured the surface finish of a metal part, obtaining the data shown. Assume that the measurements are normally distributed.

Technician1	Technician2
1.45	1.54
1.37	1.41
1.21	1.56
1.54	1.37
1.48	1.20
1.29	1.31
1.34	1.27
	1.35

(a) Test the hypothesis that the mean surface finish measurements made by the two technicians are equal. Use $\alpha = 0.05$, and assume equal variances.

Test $H_0: \mu_1 - \mu_2 = 0$ vs. $H_1: \mu_1 - \mu_2 \neq 0$. Reject H_0 if $|t_0| > t_{\alpha/2,\, n1+n2-2}$.

$$s_P = \sqrt{\frac{(n_1-1)s_1^2 + (n_2-1)s_2^2}{n_1+n_2-2}} = \sqrt{\frac{(7-1)0.115^2 + (8-1)0.125^2}{7+8-2}} = 0.1204 \qquad \text{(Equations 3-51 and 3-52)}$$

$$t_0 = \left(\bar{x}_1 - \bar{x}_2\right)\Big/\left(s_P\sqrt{1/n_1 + 1/n_2}\right) = (1.383 - 1.376)\Big/\left(0.1204\sqrt{1/7 + 1/8}\right) = 0.11$$

$t_{\alpha/2,\, n1+n2-2} = t_{0.025,\, 13} = 2.160$ (from Appendix IV)

Do not reject H_0, and conclude that there is not sufficient evidence of a difference between measurements obtained by the two technicians.

MTB>Stat>Basic Statistics>2-Sample t>Samples in different columns

```
Two-Sample T-Test and CI: Ex3-11T1, Ex3-11T2
Two-sample T for Ex3-11T1 vs Ex3-11T2
          N    Mean   StDev   SE Mean
Ex3-11T1  7   1.383   0.115    0.043
Ex3-11T2  8   1.376   0.125    0.044
Difference = mu (Ex3-11T1) - mu (Ex3-11T2)
Estimate for difference:  0.006607
95% CI for difference:  (-0.127969, 0.141183)
T-Test of difference = 0 (vs not =): T-Value = 0.11  P-Value = 0.917  DF = 13
Both use Pooled StDev = 0.1204
```

(b) What are the practical implications of the test in part (a)? Discuss what practical conclusions you would draw if the null hypothesis were rejected.

The practical implication of this test is that it does not matter which technician measures parts; the readings will be the same. If the null hypothesis had been rejected, we would have been concerned that the technicians obtained different measurements, and an investigation should be undertaken to understand why.

3-11 continued

(c) Assuming that the variances are equal, construct a 95% confidence interval on the mean difference in surface-finish measurements.

$n_1 = 7$, $\overline{x}_1 = 1.383$, $s_1 = 0.115$; $n_2 = 8$, $\overline{x}_2 = 1.376$, $s_2 = 0.125$, $s_P = 0.120$

$\alpha = 0.05$, $t_{\alpha/2,\,n1+n2-2} = t_{0.025,\,13} = 2.160$

$$(\overline{x}_1 - \overline{x}_2) - t_{\alpha/2,n_1+n_2-2}s_P\sqrt{1/n_1 + 1/n_2} \le (\mu_1 - \mu_2) \le (\overline{x}_1 - \overline{x}_2) + t_{\alpha/2,n_1+n_2-2}s_P\sqrt{1/n_1 + 1/n_2}$$

$$(1.383-1.376) - 2.1604(0.120)\sqrt{1/7 + 1/8} \le (\mu_1 - \mu_2) \le (1.383-1.376) + 2.1604(0.120)\sqrt{1/7 + 1/8}$$

$$-0.127 \le (\mu_1 - \mu_2) \le 0.141$$

(Equation 3-56)

The confidence interval for the difference contains zero. We can conclude that there is not sufficient evidence of a difference in measurements obtained by the two technicians.

(d) Test the hypothesis that the variances of the measurements made by the two technicians are equal. Use $\alpha = 0.05$. What are the practical implications if the null hypothesis is rejected?

Test $H_0 : \sigma_1^2 = \sigma_2^2$ versus $H_1 : \sigma_1^2 \ne \sigma_2^2$.

Reject H_0 if $F_0 > F_{\alpha/2,n_1-1,n_2-1}$ or $F_0 < F_{1-\alpha/2,n_1-1,n_2-1}$.

$F_0 = S_1^2 / S_2^2 = 0.115^2 / 0.125^2 = 0.8464$

$F_{\alpha/2,n_1-1,n_2-1} = F_{0.05/2,7-1,8-1} = F_{0.025,6,7} = 5.119$

$F_{1-\alpha/2,n_1-1,n_2-1} = F_{1-0.05/2,7-1,8-1} = F_{0.975,6,7} = 0.176$

MTB>Stat>Basic Statistics>2 Variances>Summarized data

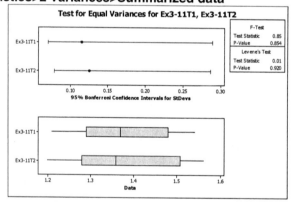

Do not reject H_0, and conclude that there is not sufficient evidence of a difference in variability between measurements obtained by the two technicians. If the null hypothesis is rejected, we would have been concerned about the difference in measurement variability between the technicians, and an investigation should be undertaken to understand why.

3-11 continued

(e) Construct a 95% confidence interval estimate of the ratio of the variances of technician measurement error.

$$\alpha = 0.05 \quad F_{1-\alpha/2,n_2-1,n_1-1} = F_{0.975,7,6} = 0.1954; \quad F_{\alpha/2,n_2-1,n_1-1} = F_{0.025,7,6} = 5.6955$$

$$\frac{S_1^2}{S_2^2} F_{1-\alpha/2,n_2-1,n_1-1} \leq \frac{\sigma_1^2}{\sigma_2^2} \leq \frac{S_1^2}{S_2^2} F_{\alpha/2,n_2-1,n_1-1}$$

$$\frac{0.115^2}{0.125^2}(0.1954) \leq \frac{\sigma_1^2}{\sigma_2^2} \leq \frac{0.115^2}{0.125^2}(5.6955)$$

$$0.165 \leq \frac{\sigma_1^2}{\sigma_2^2} \leq 4.821$$

(f) Construct a 95% confidence interval on the variance of measurement error for technician 2.

$$n_2 = 8; \quad \bar{x}_2 = 1.376; \quad S_2 = 0.125$$

$$\alpha = 0.05; \quad \chi^2_{\alpha/2,n_2-1} = \chi^2_{0.025,7} = 16.0128; \quad \chi^2_{1-\alpha/2,n_2-1} = \chi^2_{0.975,7} = 1.6899$$

$$\frac{(n-1)S^2}{\chi^2_{\alpha/2,n-1}} \leq \sigma^2 \leq \frac{(n-1)S^2}{\chi^2_{1-\alpha/2,n-1}}$$

$$\frac{(8-1)0.125^2}{16.0128} \leq \sigma^2 \leq \frac{(8-1)0.125^2}{1.6899}$$

$$0.007 \leq \sigma^2 \leq 0.065$$

(g) Does the normality assumption seem reasonable for the data?

MTB>Graph>Probability Plot>Multiple

Probability Plot of Surface Finish by Technician (Ex3-11T1, Ex3-11T2)

The normality assumption seems reasonable for these readings.

3-13. Two different hardening processes, (1) saltwater quenching and (2) oil quenching, are used on samples of a particular type of metal alloy. The results are shown here. Assume that hardness is normally distributed.

Saltwater Quench	Oil Quench
145	152
150	150
153	147
148	155
141	140
152	146
146	158
154	152
139	151
148	143

Saltwater quench: $n_1 = 10$, $\bar{x}_1 = 147.6$, $s_1 = 4.97$; Oil quench: $n_2 = 10$, $\bar{x}_2 = 149.4$, $s_2 = 5.46$

(a) Test the hypothesis that the mean hardness for the saltwater quenching process equals the mean hardness for the oil quenching process. Use $\alpha = 0.05$ and assume equal variances.

Test $H_0: \mu_1 - \mu_2 = 0$ vs. $H_1: \mu_1 - \mu_2 \neq 0$. Reject H_0 if $|t_0| > t_{\alpha/2, n1+n2-2}$.

$$s_P = \sqrt{\frac{(n_1-1)s_1^2 + (n_2-1)s_2^2}{n_1+n_2-2}} = \sqrt{\frac{(10-1)4.97^2 + (10-1)5.46^2}{10+10-2}} = 5.2217$$

(Eqns 3-51 and 3-52)

$$t_0 = (\bar{x}_1 - \bar{x}_2)/(s_P\sqrt{1/n_1 + 1/n_2}) = (147.60 - 149.40)/(5.2217\sqrt{1/10 + 1/10}) = -0.77$$

$t_{\alpha/2, n1+n2-2} = t_{0.025, 18} = 2.101$ (from Appendix IV)

Do not reject H_0, and conclude that there is not sufficient evidence of a difference between measurements produced by the two hardening processes.

MTB>Stat>Basic Statistics>2-Sample t>Samples in different columns

```
Two-Sample T-Test and CI: Ex3-13SQ, Ex3-13OQ
Two-sample T for Ex3-13SQ vs Ex3-13OQ
            N    Mean   StDev  SE Mean
Ex3-13SQ   10  147.60    4.97      1.6
Ex3-13OQ   10  149.40    5.46      1.7
Difference = mu (Ex3-13SQ) - mu (Ex3-13OQ)
Estimate for difference:  -1.80000
95% CI for difference:  (-6.70615, 3.10615)
T-Test of difference = 0 (vs not =): T-Value = -0.77  P-Value = 0.451  DF = 18
Both use Pooled StDev = 5.2217
```

(b) Assuming that the variances σ^2_1 and σ^2_2 are equal, construct a 95% confidence interval on the difference in mean hardness.

$\alpha = 0.05$, $t_{\alpha/2, n1+n2-2} = t_{0.025, 18} = 2.1009$

$$(\bar{x}_1 - \bar{x}_2) - t_{\alpha/2, n_1+n_2-2}S_P\sqrt{1/n_1 + 1/n_2} \leq (\mu_1 - \mu_2) \leq (\bar{x}_1 - \bar{x}_2) + t_{\alpha/2, n_1+n_2-2}S_P\sqrt{1/n_1 + 1/n_2}$$

$$(147.6 - 149.4) - 2.1009(5.22)\sqrt{1/10 + 1/10} \leq (\mu_1 - \mu_2) \leq (147.6 - 149.4) + 2.1009(5.22)\sqrt{1/10 + 1/10}$$

$$-6.7 \leq (\mu_1 - \mu_2) \leq 3.1$$

(Eqn 3-56)

The confidence interval for the difference contains zero. We conclude that there is not sufficient evidence of a difference between measurements produced by the two hardening processes.

3-13 continued

(c) Construct a 95% confidence interval on the ratio σ^2_1/σ^2_2. Does the assumption made earlier of equal variances seem reasonable?

$$\alpha = 0.05 \ \ F_{1-\alpha/2,n_2-1,n_1-1} = F_{0.975,9,9} = 0.2484; \ \ F_{\alpha/2,n_2-1,n_1-1} = F_{0.025,9,9} = 4.0260$$

$$\frac{S_1^2}{S_2^2}F_{1-\alpha/2,n_2-1,n_1-1} \le \frac{\sigma_1^2}{\sigma_2^2} \le \frac{S_1^2}{S_2^2}F_{\alpha/2,n_2-1,n_1-1}$$

$$\frac{4.97^2}{5.46^2}(0.2484) \le \frac{\sigma_1^2}{\sigma_2^2} \le \frac{4.97^2}{5.46^2}(4.0260)$$

$$0.21 \le \frac{\sigma_1^2}{\sigma_2^2} \le 3.34$$

Since the confidence interval includes the ratio of 1, the assumption of equal variances seems reasonable.

(d) Does the assumption of normality seem appropriate for these data?
MTB>Graph>Probability Plot>Multiple

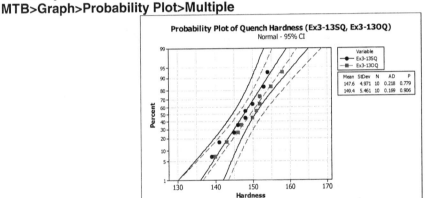

The normal distribution assumptions for both the saltwater and oil quench methods seem reasonable. However, the slopes on the normal probability plots do not appear to be the same, so the assumption of equal variances does not seem reasonable.

3-15.   A random sample of 500 connecting rod pins contains 65 nonconforming units. Estimate the process fraction nonconforming.

$n = 500$, $x = 65$, $\hat{p} = x/n = 65/500 = 0.130$

(a) Test the hypothesis that the true fraction defective in this process is 0.08. Use $\alpha = 005$.
$\quad$ $p_0 = 0.08$, $\alpha = 0.05$
$\quad$ Test H_0: $p = 0.08$ versus H_1: $p \neq 0.08$. Reject H_0 if $|Z_0| > Z_{\alpha/2}$.
$\quad$ $np_0 = 500(0.08) = 40$
$\quad$ Since $(x = 65) > (np_0 = 40)$, use the normal approximation to the binomial for $x > np_0$.
$$Z_0 = \frac{(x-0.5) - np_0}{\sqrt{np_0(1-p_0)}} = \frac{(65-0.5) - 40}{\sqrt{40(1-0.08)}} = 4.0387$$
$\quad$ $Z_{\alpha/2} = Z_{0.05/2} = Z_{0.025} = 1.96$
$\quad$ Reject H_0, and conclude the sample process fraction nonconforming differs from 0.08.
$\quad$ **MTB>Stat>Basic Statistics>1 Proportion>Summarized data**

Test and CI for One Proportion
```
Test of p = 0.08 vs p not = 0.08
Sample    X    N   Sample p          95% CI         Z-Value   P-Value
1        65   500  0.130000   (0.100522, 0.159478)    4.12     0.000
```
$\quad$ Note that MINITAB uses an exact method, not an approximation.

(b) Find the *P*-value for this test.
$\quad$ Find the corresponding cumulative standard normal, $\Phi(Z_0)$ for $Z_0 = 4.0387$, and calculate:
$\quad$ P-value $= 2[1 - \Phi|Z_0|] = 2[1 - \Phi|4.0387|] = 2[1 - 0.99997] = 0.00006$

(c) Construct a 95% upper confidence interval on the true process fraction nonconforming.
$\quad$ $\alpha = 0.05$, $Z_\alpha = Z_{0.05} = 1.645$
$$p \leq \hat{p} + Z_\alpha\sqrt{\hat{p}(1-\hat{p})/n}$$
$$p \leq 0.13 + 1.645\sqrt{0.13(1-0.13)/500}$$
$$p \leq 0.155$$
$\quad$ The 95% upper confidence interval for the process fraction nonconforming is 0.155.

3-17. A new purification unit is installed in a chemical process. Before its installation, a random sample yielded the following data about the percentage of impurity: $\bar{x}_1 = 9.85$, $s_1^2 = 6.79$, and $n_1 = 10$. After installation, a random sample resulted in $\bar{x}_2 = 8.08$, $s_2^2 = 6.18$, and $n_2 = 8$.

(a) Can you conclude that the two variances are equal? Use $\alpha = 0.05$.

Test $H_0 : \sigma_1^2 = \sigma_2^2$ versus $H_1 : \sigma_1^2 \neq \sigma_2^2$, at $\alpha = 0.05$

Reject H_0 if $F_0 > F_{\alpha/2, n_1-1, n_2-2}$ or $F_0 < F_{1-\alpha/2, n_1-1, n_2-1}$

$F_{\alpha/2, n_1-1, n_2-2} = F_{0.025, 9, 7} = 4.8232;\quad F_{1-\alpha/2, n_1-1, n_2-1} = F_{0.975, 9, 7} = 0.2383$

$F_0 = S_1^2 / S_2^2 = 6.79/6.18 = 1.0987$

$F_0 = 1.0987 < 4.8232$ and > 0.2383, so do not reject H_0

MTB>Stat>Basic Statistics>2 Variances>Summarized data

```
Test for Equal Variances
95% Bonferroni confidence intervals for standard deviations
Sample   N    Lower    StDev    Upper
   1    10  1.70449  2.60576  5.24710
   2     8  1.55525  2.48596  5.69405
F-Test (normal distribution)
Test statistic = 1.10, p-value = 0.922
```

The impurity variances before and after installation are the same.

(b) Can you conclude that the new purification device has reduced the mean percentage of impurity? Use $\alpha = 0.05$.

Test $H_0: \mu_1 = \mu_2$ versus $H_1: \mu_1 > \mu_2$, $\alpha = 0.05$.

Reject H_0 if $t_0 > t_{\alpha, n1+n2-2}$.

$t_{\alpha, n1+n2-2} = t_{0.05, 10+8-2} = 1.746$

$$S_P = \sqrt{\frac{(n_1-1)S_1^2 + (n_2-1)S_2^2}{n_1+n_2-2}} = \sqrt{\frac{(10-1)6.79 + (8-1)6.18}{10+8-2}} = 2.554$$

$$t_0 = \frac{\bar{x}_1 - \bar{x}_2}{S_P\sqrt{1/n_1 + 1/n_2}} = \frac{9.85 - 8.08}{2.554\sqrt{1/10 + 1/8}} = 1.461$$

MTB>Stat>Basic Statistics>2-Sample t>Summarized data

```
Two-Sample T-Test and CI
Sample   N   Mean   StDev   SE Mean
1       10   9.85   2.61     0.83
2        8   8.08   2.49     0.88
Difference = mu (1) - mu (2)
Estimate for difference:  1.77000
95% lower bound for difference:  -0.34856
T-Test of difference = 0 (vs >): T-Value = 1.46  P-Value = 0.082  DF = 16
Both use Pooled StDev = 2.5582
```

The mean impurity after installation of the new purification unit is not less than before.

3-19.   The diameter of a metal rod is measured by 12 inspectors, each using both a micrometer caliper and a vernier caliper. The results are shown here. Is there a difference between the mean measurements produced by the two types of caliper? Use $\alpha = 0.01$.

Inspector	Micrometer Caliper	Vernier Caliper
1	0.150	0.151
2	0.151	0.150
3	0.151	0.151
4	0.152	0.150
5	0.151	0.151
6	0.150	0.151
7	0.151	0.153
8	0.153	0.155
9	0.152	0.154
10	0.151	0.151
11	0.151	0.150
12	0.151	0.152

Test H_0: $\mu_d = 0$ versus H_1: $\mu_d \neq 0$. Reject H_0 if $|t_0| > t_{\alpha/2,\, n1 + n2 - 2}$.

$$\bar{d} = \frac{1}{n} \sum_{j=1}^{n} \left(x_{\text{Micrometer},j} - x_{\text{Vernier},j} \right) = \frac{1}{12}\left[(0.150 - 0.151) + \cdots + (0.151 - 0.152) \right] = -0.000417$$

$$S_d^2 = \frac{\sum_{j=1}^{n} d_j^2 - \left(\sum_{j=1}^{n} d_j \right)^2 \Big/ n}{(n-1)} = 0.001311^2$$

$$t_0 = \bar{d} \Big/ \left(S_d / \sqrt{n} \right) = -0.000417 \Big/ \left(0.001311 / \sqrt{12} \right) = -1.10$$

$$t_{\alpha/2,\, n1 + n2 - 2} = t_{0.005,22} = 2.8188$$

$(|t_0| = 1.10) < 2.8188$, so do not reject H_0. There is no strong evidence to indicate that the two calipers differ in their mean measurements.

MTB>Stat>Basic Statistics>Paired t>Samples in Columns

```
Paired T-Test and CI: Ex3-19MC, Ex3-19VC
Paired T for Ex3-19MC - Ex3-19VC
              N       Mean      StDev     SE Mean
Ex3-19MC      12   0.151167   0.000835   0.000241
Ex3-19VC      12   0.151583   0.001621   0.000468
Difference    12  -0.000417   0.001311   0.000379
95% CI for mean difference: (-0.001250, 0.000417)
T-Test of mean difference = 0 (vs not = 0): T-Value = -1.10   P-Value = 0.295
```

3-21.   An experiment was conducted to investigate the filling capability of packaging equipment at a winery in Newberg, Oregon. Twenty bottles of Pinot Gris were randomly selected and the fill volume (in ml) measured. Assume that fill volume has a normal distribution. The data are as follows: 753, 751, 752, 753, 753, 753, 752, 753, 754, 754, 752, 751, 752, 750, 753, 755, 753, 756, 751, and 750.

(a) Do the data support the claim that the standard deviation of fill volume is less than 1 ml? Use $\alpha = 0.05$.

$n = 20$, $\bar{x} = 752.6$ ml, $s = 1.5$, $\alpha = 0.05$

Test H_0: $\sigma^2 = 1$ versus H_1: $\sigma^2 < 1$. Reject H_0 if $\chi_0^2 < \chi_{1-\alpha, n-1}^2$.

$\chi_0^2 = \left[(n-1)S^2 \right]/\sigma_0^2 = \left[(20-1)1.5^2 \right]/1 = 42.75$

$\chi_{1-\alpha, n-1}^2 = \chi_{0.95,19}^2 = 10.1170$ (from Appendix III)

$(\chi_0^2 = 42.75) > 10.1170$, so do not reject H_0. The standard deviation of the fill volume is not less than 1ml.

(b) Find a 95% two-sided confidence interval on the standard deviation of fill volume.

$\chi_{\alpha/2, n-1}^2 = \chi_{0.025,19}^2 = 32.85$; $\chi_{1-\alpha/2, n-1}^2 = \chi_{0.975,19}^2 = 8.91$ (from Appendix III)

$(n-1)S^2 / \chi_{\alpha/2,n-1}^2 \le \sigma^2 \le (n-1)S^2 / \chi_{1-\alpha/2,n-1}^2$

$(20-1)1.5^2 / 32.85 \le \sigma^2 \le (20-1)1.5^2 / 8.91$

$$1.30 \le \sigma^2 \le 4.80$$

$$1.14 \le \sigma \le 2.19$$

Also, in MINITAB: **MTB>Stat>Basic Statistics>Graphical Summary**

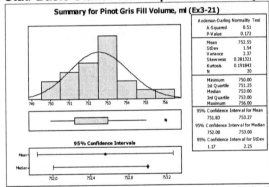

The confidence interval does not include unity; therefore, we cannot conclude that the standard deviation of fill volume is less than 1 ml.

(c) Does it seem reasonable to assume that fill volume has a normal distribution?

MTB>Graph>Probability Plot>Single

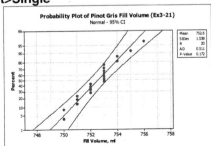

The plotted points do not fall approximately along a straight line, so the assumption that battery life is normally distributed is not appropriate.

3-23. Consider the hypotheses

$$H_0: \quad \mu = \mu_0$$
$$H_1: \quad \mu \neq \mu_0$$

where σ^2 is known. Derive a general expression for determining the sample size for detecting a true mean of $\mu_1 \neq \mu_0$ with probability $1 - \beta$ if the type I error rate is α.

Let $\mu_1 = \mu_0 + \delta$.

From equation. 3-46, $\beta = \Phi\left(Z_{\alpha/2} - \delta\sqrt{n}/\sigma\right) - \Phi\left(-Z_{\alpha/2} - \delta\sqrt{n}/\sigma\right)$

If $\delta > 0$, then $\Phi\left(-Z_{\alpha/2} - \delta\sqrt{n}/\sigma\right)$ is likely to be small compared with β. So,

$$\beta \approx \Phi\left(Z_{\alpha/2} - \delta\sqrt{n}/\sigma\right)$$
$$\Phi(\beta) \approx \Phi^{-1}\left(Z_{\alpha/2} - \delta\sqrt{n}/\sigma\right)$$
$$-Z_\beta \approx Z_{\alpha/2} - \delta\sqrt{n}/\sigma$$
$$n \approx \left[(Z_{\alpha/2} + Z_\beta)\sigma/\delta\right]^2$$

3-27. An inspector counts the surface-finish defects in dishwashers. A random sample of five dishwashers contains three such defects. Is there reason to conclude that the mean occurrence rate of surface finish defects per dishwasher exceeds 0.5? Use the results of part (a) of Exercise 3-26 and assume that $\alpha = 0.05$.

$x \sim \text{Poi}(\lambda)$, $n = 5$, $x = 3$, $\overline{x} = x/N = 3/5 = 0.6$

Test H_0: $\lambda = 0.5$ versus H_1: $\lambda > 0.5$, at $\alpha = 0.05$. Reject H_0 if $Z_0 > Z_\alpha$.

$Z_\alpha = Z_{0.05} = 1.645$

$Z_0 = \left(\overline{x} - \lambda_0\right)/\sqrt{\lambda_0/n} = (0.6 - 0.5)/\sqrt{0.5/5} = 0.3162$

$(Z_0 = 0.3162) < 1.645$, so do not reject H_0.

3-29. An article in *Solid State Technology* (May 1987) describes an experiment to determine the effect of C_2F_6 flow rate on etch uniformity on a silicon wafer used in integrated-circuit manufacturing. Three flow rates are tested, and the resulting uniformity (in percent) is observed for six test units at each flow rate. The data are shown in the following table.

C_2F_6 Flow (SCCM)	Observations					
	1	2	3	4	5	6
125	2.7	2.6	4.6	3.2	3.0	3.8
160	4.6	4.9	5.0	4.2	3.6	4.2
200	4.6	2.9	3.4	3.5	4.1	5.1

(a) Does C_2F_6 flow rate affect etch uniformity? Answer this question by using an analysis of variance with $\alpha = 0.05$.

MTB>Stat>ANOVA>One-Way

```
One-way ANOVA: Ex3-29Obs versus Ex3-29Flow
Source       DF      SS      MS      F       P
Ex3-29Flow    2    3.648   1.824   3.59   0.053
Error        15    7.630   0.509
Total        17   11.278
S = 0.7132    R-Sq = 32.34%    R-Sq(adj) = 23.32%
                      Individual 95% CIs For Mean Based on
                      Pooled StDev
Level  N    Mean    StDev    -----+---------+---------+---------+----
125    6   3.3167   0.7600   (---------*----------)
160    6   4.4167   0.5231                    (----------*---------)
200    6   3.9333   0.8214             (----------*---------)
                             -----+---------+---------+---------+----
                              3.00      3.60      4.20      4.80

Pooled StDev = 0.7132
```

$(F_{0.05,2,15} = 3.6823) > (F_0 = 3.59)$, so flow rate does not affect etch uniformity at a significance level $\alpha = 0.05$. However, the P-value = 0.053 is just slightly greater than 0.05, so there is some evidence that gas flow rate affects the etch uniformity.

(b) Construct a box plot of the etch uniformity data. Use this plot, together with the analysis of variance results, to determine which gas flow rate would be best in terms of etch uniformity (a small percentage is best).

MTB>Stat>ANOVA>One-Way>Graphs, Boxplots of data (also MTB>Graph>Boxplot>One Y, With Groups)

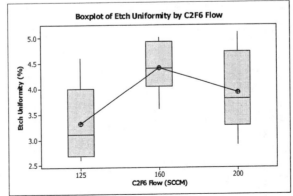

Gas flow rate of 125 SCCM gives smallest mean percentage uniformity.

3-29 continued

(c) Plot the residuals versus predicted C_2F_6 flow. Interpret this plot.

MTB>Stat>ANOVA>One-Way>Graphs, Residuals versus fits

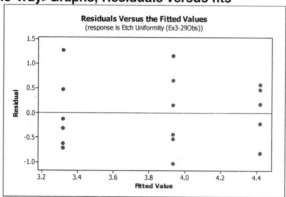

Residuals are satisfactory, with no unusual patterns and approximately equal variance over the range of fitted values.

(d) Does the normality assumption seem reasonable in this problem?

MTB>Stat>ANOVA>One-Way>Graphs, Normal plot of residuals

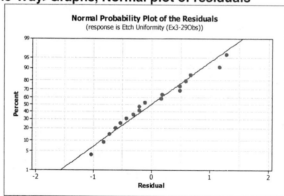

The normality assumption is reasonable.

3-31.   An article in the *ACI Materials Journal* (Vol. 84, 1987, pp. 213–216) describes several experiments investigating the rodding of concrete to remove entrapped air. A 3-in. diameter cylinder was used, and the number of times this rod was used is the design variable. The resulting compressive strength of the concrete specimen is the response. The data are shown in the following table.

Rodding Level	Compressive Strength		
10	1530	1530	1440
15	1610	1650	1500
20	1560	1730	1530
25	1500	1490	1510

(a) Is there any difference in compressive strength due to the rodding level? Answer this question by using the analysis of variance with $\alpha = 0.05$.

MTB>Stat>ANOVA>One-Way>Graphs>Boxplots of data, Normal plot of residuals

```
One-way ANOVA: Ex3-31Str versus Ex3-31Rod
Source      DF    SS     MS     F      P
Ex3-31Rod    3   28633  9544  1.87  0.214
Error        8   40933  5117
Total       11   69567
S = 71.53   R-Sq = 41.16%   R-Sq(adj) = 19.09%
                           Individual 95% CIs For Mean Based on
                           Pooled StDev
Level   N    Mean   StDev   ----+---------+---------+---------+-----
10      3   1500.0   52.0   (-----------*----------)
15      3   1586.7   77.7              (-----------*-----------)
20      3   1606.7  107.9                  (-----------*-----------)
25      3   1500.0   10.0   (-----------*----------)
                           ----+---------+---------+---------+-----
                           1440       1520      1600      1680
Pooled StDev = 71.5
```

ANOVA *P*-value = 0.214, so no difference due to rodding level at $\alpha = 0.05$.

(b) Construct box plots of compressive strength by rodding level. Provide a practical interpretation of these plots.

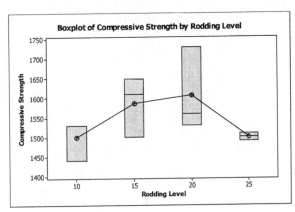

Level 25 exhibits considerably less variability than the other three levels.

3-31 continued

(c) Construct a normal probability plot of the residuals from this experiment. Does the assumption of a normal distribution for compressive strength seem reasonable?

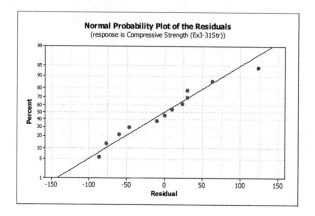

The normal distribution assumption for compressive strength is reasonable.

3-33.   An aluminum producer manufactures carbon anodes and bakes them in a ring furnace prior to use in the smelting operation. The baked density of the anode is an important quality characteristic, as it may affect anode life. One of the process engineers suspects that firing temperature in the ring furnace affects baked anode density. An experiment was run at four different temperature levels, and six anodes were baked at each temperature level. The data from the experiment follow.

Temperature	(°C)Density					
500	41.8	41.9	41.7	41.6	41.5	41.7
525	41.4	41.3	41.7	41.6	41.7	41.8
550	41.2	41.0	41.6	41.9	41.7	41.3
575	41.0	40.6	41.8	41.2	41.9	41.5

(a) Does firing temperature in the ring furnace affect mean baked anode density?

MTB>Stat>ANOVA>One-Way>Graphs>Boxplots of data, Normal plot of residuals

```
One-way ANOVA: Ex3-33Den versus Ex3-33T
Source    DF      SS      MS      F      P
Ex3-33T    3   0.457   0.152   1.45   0.258
Error     20   2.097   0.105
Total     23   2.553
S = 0.3238    R-Sq = 17.89%    R-Sq(adj) = 5.57%
                                 Individual 95% CIs For Mean Based on
                                 Pooled StDev
Level   N    Mean    StDev   --------+---------+---------+---------+-
500     6   41.700   0.141                   (----------*----------)
525     6   41.583   0.194              (----------*----------)
550     6   41.450   0.339          (----------*----------)
575     6   41.333   0.497   (----------*----------)
                             --------+---------+---------+---------+-
                              41.25     41.50     41.75     42.00
Pooled StDev = 0.324
```

ANOVA *P*-value = 0.258. At $\alpha = 0.05$, temperature level does not significantly affect mean baked anode density.

3-33 continued

(b) Find the residuals for this experiment and plot them on a normal probability scale. Comment on the plot.

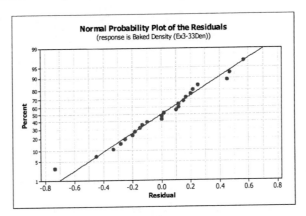

Normality assumption is reasonable.

(c) What firing temperature would you recommend using?

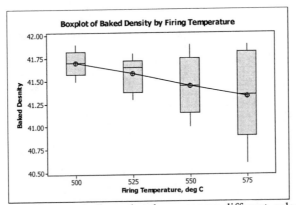

Since statistically there is no evidence to indicate that the means are different, select the temperature with the smallest variance, 500°C (see Boxplot), which probably also incurs the smallest cost (lowest temperature).

3-35.   An article in *Environmental International* (Vol. 18, No. 4, 1992) describes an experiment in which the amount of radon released in showers was investigated. Radon-enriched water was used in the experiment, and six different orifice diameters were tested in shower heads. The data from the experiment are shown in the following table.

Orifice Diameter	Radon Released (%)			
0.37	80	83	83	85
0.51	75	75	79	79
0.71	74	73	76	77
1.02	67	72	74	74
1.40	62	62	67	69
1.99	60	61	64	66

(a) Does the size of the orifice affect the mean percentage of radon released? Use the analysis of variance and $\alpha = 0.05$.

MTB>Stat>ANOVA>One-Way>Graphs>Boxplots of data

```
One-way ANOVA: Ex3-35Rad versus Ex3-35Dia
Source      DF        SS       MS       F       P
Ex3-35Dia    5   1133.38   226.68   30.85   0.000
Error       18    132.25     7.35
Total       23   1265.63
S = 2.711   R-Sq = 89.55%   R-Sq(adj) = 86.65%
                            Individual 95% CIs For Mean Based on
                            Pooled StDev
Level   N    Mean   StDev   ----+---------+---------+---------+-----
0.37    4  82.750   2.062                              (---*---)
0.51    4  77.000   2.309                     (---*---)
0.71    4  75.000   1.826                 (---*---)
1.02    4  71.750   3.304            (----*---)
1.40    4  65.000   3.559    (---*---)
1.99    4  62.750   2.754  (---*---)
                            ----+---------+---------+---------+-----
                            63.0      70.0      77.0      84.0
Pooled StDev = 2.711
```

Orifice size does affect mean % radon release, at $\alpha = 0.05$.

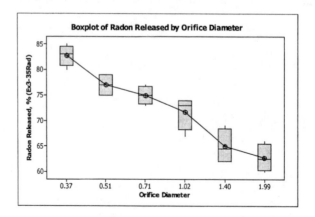

Smallest % radon released at 1.99 and 1.4 orifice diameters.

3-35 continued

(b) Analyze the results from this experiment.

MTB>Stat>ANOVA>One-Way>Graphs>Normal plot of residuals, Residuals versus fits, Residuals versus the Variables

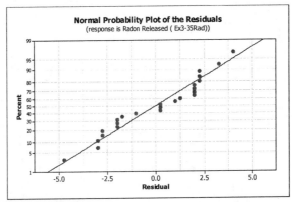

Residuals violate the normality distribution.

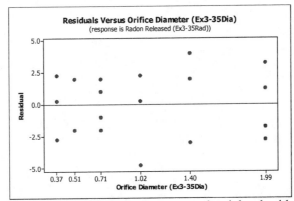

The assumption of equal variance at each factor level appears to be violated, with larger variances at the larger diameters (1.02, 1.40, 1.99).

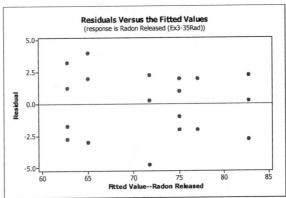

Variability in residuals does not appear to depend on the magnitude of predicted (or fitted) values.

CHAPTER 4

Methods and Philosophy of Statistical Process Control

Learning Objectives

After completing this chapter you should be able to:

1. Understand chance and assignable causes of variability in a process
2. Explain the statistical basis of the Shewhart control chart, including choice of sample size, control limits, and sampling interval
3. Explain the rational subgroup concept
4. Understand the basic tools of SPC; the histogram or stem-and-leaf plot, the check sheet, the Pareto chart, the cause-and-effect diagram, the defect concentration diagram, the scatter diagram, and the control chart
5. Explain phase I and phase II use of control charts
6. Explain how average run length is used as a performance measure for a control chart
7. Explain how sensitizing rules and pattern recognition are used in conjunction with control charts

Important Terms and Concepts

Assignable causes of variation
Average time to signal
Chance causes of variation
Control limits
Designed experiments
Histogram
"Magnificent seven"
Out-of-control process
Patterns on a control charts
Rational subgroups
Sampling frequency for control charts
Sensitizing rules for control charts
Statistical control of a process
Stem-and-leaf plot
Warning limits

Average run length (ARL)
Cause-and-effect diagram
Control chart
Defect concentration diagram
Flow charts and operations process charts
In-control process
Out-of-control-action plan (OCAP)
Pareto chart
Phase I and phase II application of control charts
Sample size for control charts
Scatter diagram
Shewhart control charts
Statistical process control (SPC)
Three-sigma control limits

Exercises

4-17. Consider the control chart shown here. Does the pattern appear random?

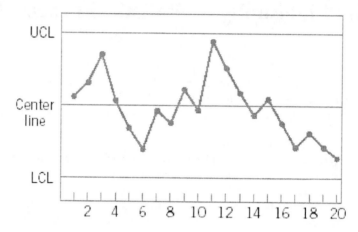

Evidence of runs, trends or cycles? NO. There are no runs of 5 points or cycles. So, we can say that the plot point pattern appears to be random.

4-21. Apply the Western Electric rules to the control chart in Exercise 4-17. Are any of the criteria for declaring the process out of control satisfied?

Check:
- Any point outside the 3-sigma control limits? NO.
- 2 of 3 beyond 2 sigma of centerline? NO.
- 4 of 5 at 1 sigma or beyond of centerline? YES. Points #17, 18, 19, and 20 are outside the lower 1-sigma area.
- 8 consecutive points on one side of centerline? NO.

A one out-of-control criterion is satisfied.

4-19.   Consider the control chart shown here. Does the pattern appear random?

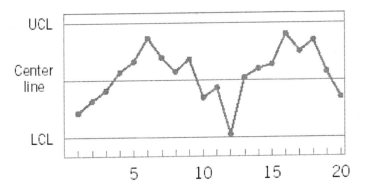

Evidence of runs, trends or cycles? YES, there is a "low - high - low - high - low" wave (all samples), which might be a cycle. So, we can say that the pattern does not appear random.

4-23.   Apply the Western Electric rules to the control chart presented in Exercise 4-19. Would these rules result in any out-of-control signals?
   Check:
   - Any point outside the 3-sigma control limits? NO. (Point #12 is within the lower 3-sigma control limit.)
   - 2 of 3 beyond 2 sigma of centerline? YES, points #16, 17, and 18.
   - 4 of 5 at 1 sigma or beyond of centerline? YES, points #5, 6, 7, 8, and 9.
   - 8 consecutive points on one side of centerline? NO.
   Two out-of-control criteria are satisfied.

CHAPTER 5

Control Charts for Variables

Learning Objectives

After completing this chapter you should be able to:

1. Understand the statistical basis of Shewhart control charts for variables
2. Know how to design variables control charts
3. Know how to set up and use $\bar{x}$ and R control charts
4. Know how to estimate process capability from the control chart information
5. Know how to interpret patterns on $\bar{x}$ and R control charts
6. Know how to set up and use $\bar{x}$ and s or s^2 control charts
7. Know how to set up and use control charts for individual measurements
8. Understand the importance of the normality assumption for individuals control charts and know how to check this information
9. Understand the rational subgroup concept for variables control charts
10. Determine the average run length for variables control charts

Important Terms and Concepts

Average run length
Interpretation of control charts
Natural tolerance limits for a process
Operating characteristic (OC) curve for the $\bar{x}$ control chart
Phase I control chart usage
Probability limits for control charts
Process capability ratio (PCR) Cp
Rational subgroups
s^2 control chart
Specification limits
Tier chart or tolerance diagram
Variable sample size on control charts
$\bar{x}$ control chart

Control chart for individuals
Moving-range control chart
Normality and control charts
Patterns on control charts
Phase II control chart usage
Process capability
R control chart
s control chart
Shewhart control charts
Three-sigma control limits
Trial control limits
Variables control charts

Exercises

Notes:
1. The MINITAB convention for determining whether a point is out of control is: (1) if a plot point is within the control limits, it is in control, or (2) if a plot point is on or beyond the limits, it is out of control.
2. MINITAB uses pooled standard deviation to estimate standard deviation for control chart limits and capability estimates. This can be changed in dialog boxes or under **Tools>Options>Control Charts and Quality Tools>Estimating Standard Deviation**.
3. MINITAB defines some sensitizing rules for control charts differently than the standard rules. In particular, a run of n consecutive points on one side of the center line is defined as 9 points, not 8. This can be changed under **Tools>Options>Control Charts and Quality Tools>Define Tests**.

5-1. The data shown here are $\bar{x}$ and R values for 24 samples of size $n = 5$ taken from a process producing bearings. The measurements are made on the inside diameter of the bearing, with only the last three decimals recorded (i.e., 34.5 should be 0.50345).

Sample Number	$\bar{x}$	R	Sample Number	$\bar{x}$	R
1	34.5	3	13	35.4	8
2	34.2	4	14	34.0	6
3	31.6	4	15	37.1	5
4	31.5	4	16	34.9	7
5	35.0	5	17	33.5	4
6	34.1	6	18	31.7	3
7	32.6	4	19	34.0	8
8	33.8	3	20	35.1	4
9	34.8	7	21	33.7	2
10	33.6	8	22	32.8	1
11	31.9	3	23	33.5	3
12	38.6	9	24	34.2	2

(a) Set up $\bar{x}$ and R charts on this process. Does the process seem to be in statistical control? If necessary, revise the trial control limits.

Note: MINITAB does not include functionality for constructing control charts from summary statistics (such as $\bar{x}$ and R). Excel can be used to calculate control limits and to construct control charts.

For $n = 5$, $A_2 = 0.577$, $D_4 = 2.114$, $D_3 = 0$ (from Appendix VI)

$$\bar{\bar{x}} = \frac{\bar{x}_1 + \bar{x}_2 + \cdots + \bar{x}_m}{m} = \frac{34.5 + 34.2 + \cdots + 34.2}{24} = 34.00$$

$$\bar{R} = \frac{R_1 + R_2 + \cdots + R_m}{m} = \frac{3 + 4 + \cdots + 2}{24} = 4.71$$

$\text{UCL}_{\bar{x}} = \bar{\bar{x}} + A_2\bar{R} = 34.00 + 0.577(4.71) = 36.72$

$\text{CL}_{\bar{x}} = \bar{\bar{x}} = 34.00$ (Equations 5-2, -3, -4 and -5)

$\text{LCL}_{\bar{x}} = \bar{\bar{x}} - A_2\bar{R} = 34.00 - 0.577(4.71) = 31.29$

$\text{UCL}_R = D_4\bar{R} = 2.115(4.71) = 9.96$

$\text{CL}_R = \bar{R} = 4.71$

$\text{LCL}_R = D_3\bar{R} = 0(4.71) = 0.00$

5-1(a) continued

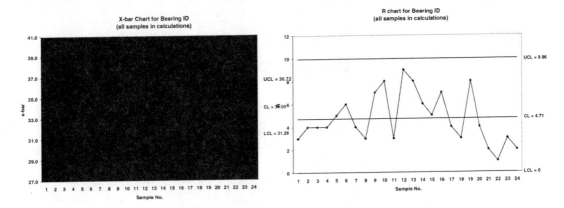

The process is not in statistical control; $\bar{x}$ is beyond the upper control limit for both Sample No. 12 and Sample No. 15. Assuming an assignable cause is found for these two out-of-control points, the two samples can be excluded from the control limit calculations. The new process parameter estimates are:

$\bar{\bar{x}} = 33.65; \quad \bar{R} = 4.5; \quad \hat{\sigma}_x = \bar{R}/d_2 = 4.5/2.326 = 1.93$

$UCL_{\bar{x}} = 36.25; CL_{\bar{x}} = 33.65; LCL_{\bar{x}} = 31.06$ \qquad (Equations 5-2, -3, -4, -5 and -6)

$UCL_R = 9.52; CL_R = 4.5; LCL_R = 0.00$

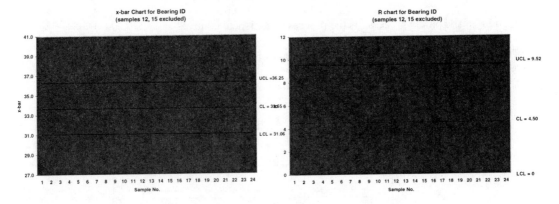

(b) If specifications on this diameter are 0.5030 ± 0.0010, find the percentage of nonconforming bearings produced by this process. Assume that diameter is normally distributed.

$$\hat{p} = Pr\{x < LSL\} + Pr\{x > USL\} = Pr\{x < 20\} + Pr\{x > 40\} = Pr\{x < 20\} + \left[1 - Pr\{x < 40\}\right]$$

$$= \Phi\left(\frac{20 - 33.65}{1.93}\right) + \left[1 - \Phi\left(\frac{40 - 33.65}{1.93}\right)\right] \qquad \text{(text p. 202)}$$

$$= \Phi(-7.07) + 1 - \Phi(3.29) = 0 + 1 - 0.99950 = 0.00050$$

5-3.   The data shown here are the deviations from nominal diameter for holes drilled in a carbon-fiber composite material used in aerospace manufacturing. The values reported are deviations from nominal in ten-thousandths of an inch.

Sample Number	x_1	x_2	x_3	x_4	x_5
1	−30	+50	−20	+10	+30
2	0	+50	−60	−20	+30
3	−50	+10	+20	+30	+20
4	−10	−10	+30	−20	+50
5	+20	−40	+50	+20	+10
6	0	0	+40	−40	+20
7	0	0	+20	−20	−10
8	+70	−30	+30	−10	0
9	0	0	+20	−20	+10
10	+10	+20	+30	+10	+50
11	+40	0	+20	0	+20
12	+30	+20	+30	+10	+40
13	+30	−30	0	+10	+10
14	+30	−10	+50	−10	−30
15	+10	−10	+50	+40	0
16	0	0	+30	−10	0
17	+20	+20	+30	+30	−20
18	+10	−20	+50	+30	+10
19	+50	−10	+40	+20	0
20	+50	0	0	+30	+10

(a) Set up $\bar{x}$ and R charts on the process. Is the process in statistical control?

MTB>Stat>Control Charts>Variables Charts for Subgroups>Xbar-R (Ex5-3Dia)

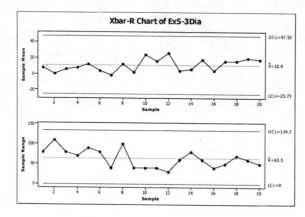

$\bar{x}$ chart: UCL = 47.53, CL = 10.9, LCL = 25.73
R chart: UCL = 134.3, CL = 63.5, LCL = 0

The process is in statistical control with no out-of-control signals, runs, trends, or cycles.

5-3 continued

(b) Estimate the process standard deviation using the range method.

$$\hat{\sigma}_x = \bar{R} / d_2 = 63.5 / 2.326 = 27.3$$

(c) If specifications are at nominal ± 100, what can you say about the capability of this process? Calculate the PCR C_p.

USL = +100, LSL = −100

$$\hat{C}_P = \frac{USL - LSL}{6\hat{\sigma}_x} = \frac{+100 - (-100)}{6(27.3)} = 1.22 \text{, so the process is capable.}$$

MTB>Stat>Quality Tools>Capability Analysis>Normal (Ex5-3Dia)

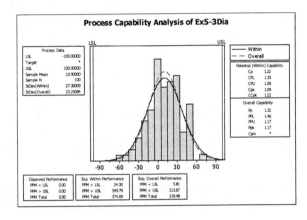

5-5.   The fill volume of soft-drink beverage bottles is an important quality characteristic. The volume is measured (approximately) by placing a gauge over the crown and comparing the height of the liquid in the neck of the bottle against a coded scale. On this scale, a reading of zero corresponds to the correct fill height. Fifteen samples of size $n = 10$ have been analyzed, and the fill heights are shown next.

Sample Number	x_1	x_2	x_3	x_4	x_5	x_6	x_7	x_8	x_9	x_{10}
1	2.5	0.5	2.0	-1.0	1.0	-1.0	0.5	1.5	0.5	-1.5
2	0.0	0.0	0.5	1.0	1.5	1.0	-1.0	1.0	1.5	-1.0
3	1.5	1.0	1.0	-1.0	0.0	-1.5	-1.0	-1.0	1.0	-1.0
4	0.0	0.5	-2.0	0.0	-1.0	1.5	-1.5	0.0	-2.0	-1.5
5	0.0	0.0	0.0	-0.5	0.5	1.0	-0.5	-0.5	0.0	0.0
6	1.0	-0.5	0.0	0.0	0.0	0.5	-1.0	1.0	-2.0	1.0
7	1.0	-1.0	-1.0	-1.0	0.0	1.5	0.0	1.0	0.0	0.0
8	0.0	-1.5	-0.5	1.5	0.0	0.0	0.0	-1.0	0.5	-0.5
9	-2.0	-1.5	1.5	1.5	0.0	0.0	0.5	1.0	0.0	1.0
10	-0.5	3.5	0.0	-1.0	-1.5	-1.5	-1.0	-1.0	1.0	0.5
11	0.0	1.5	0.0	0.0	2.0	-1.5	0.5	-0.5	2.0	-1.0
12	0.0	-2.0	-0.5	0.0	-0.5	2.0	1.5	0.0	0.5	-1.0
13	-1.0	-0.5	-0.5	-1.0	0.0	0.5	0.5	-1.5	-1.0	-1.0
14	0.5	1.0	-1.0	-0.5	-2.0	-1.0	-1.5	0.0	1.5	1.5
15	1.0	0.0	1.5	1.5	1.0	-1.0	0.0	1.0	-2.0	-1.5

(a) Set up $\bar{x}$ and s control charts on this process. Does the process exhibit statistical control? If necessary, construct revised control limits.

MTB>Stat>Control Charts>Variables Charts for Subgroups>Xbar-S (Ex5-5Vol)
Under "Options, Estimate" select Sbar as method to estimate standard deviation.

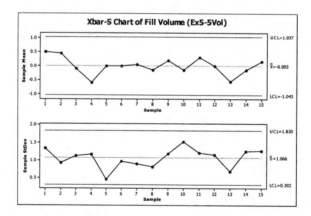

$\bar{x}$ chart: UCL = 1.037, CL = 0.003, LCL = -1.043
s chart: UCL = 1.830, CL = 1.066, LCL = 0.302

The process is in statistical control, with no out-of-control signals, runs, trends, or cycles.

5-5 continued

(b) Set up an R chart, and compare with the s chart in part (a).
   MTB>Stat>Control Charts>Variables Charts for Subgroups>Xbar-R (Ex5-5Vol)

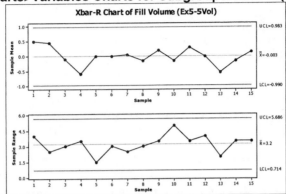

The process is in statistical control, with no out-of-control signals, runs, trends, or cycles. There is no difference in interpretation from the $\bar{x} - s$ chart.

(c) Set up an s^2 chart, and compare with the s chart in part (a).
   Let $\alpha = 0.010$. $n = 15$, $\bar{s} = 1.066$.
   $$CL = \bar{s}^2 = 1.066^2 = 1.136$$
   $$UCL = \bar{s}^2/(n-1)\,\chi^2_{\alpha/2,n-1} = 1.066^2/(15-1)\left(\chi^2_{0.010/2,15-1}\right) = 1.066^2/(15-1)(31.32) = 2.542$$
   $$LCL = \bar{s}^2/(n-1)\,\chi^2_{1-(\alpha/2),n-1} = 1.066^2/(15-1)\left(\chi^2_{1-(0.010/2),15-1}\right) = 1.066^2/(15-1)(4.07) = 0.330$$

MINITAB does not have an s^2 or variance chart. To construct an s^2 control chart, first calculate the sample standard deviations and then create a time series plot. To obtain sample standard deviations: **Stat>Basic Statistics>Store Descriptive Statistics**. "Variables" is column with sample data (Ex5-5Vol), and "By Variables" is the sample ID column (Ex5-5Sample). In "Statistics" select "Variance". Results are displayed in the session window. Copy results from the session window by holding down the keyboard "Alt" key, selecting only the variance column, and then copying & pasting to an empty worksheet column (results in Ex5-5Variance).

Graph>Time Series Plot>Simple (Add limits using: Time/Scale>Reference Lines>Y positions)

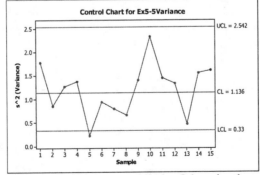

Sample 5 signals out of control below the lower control limit. Otherwise there are no runs, trends, or cycles. If the limits had been calculated using $\alpha = 0.0027$ (not tabulated in textbook), sample 5 would be within the limits, and there would be no difference in interpretation from either the $\bar{x} - s$ or the $x-R$ chart.

5-7. Rework Exercise 5-2 using the s chart.

$$\overline{\overline{x}} = \frac{\overline{x}_1 + \overline{x}_2 + \cdots + \overline{x}_{20}}{20} = 10.33$$

$$\overline{s} = \frac{s_1 + s_2 + \cdots + s_{20}}{20} = 2.703$$

$$\text{UCL}_{\overline{x}} = \overline{\overline{x}} + A_3\overline{s} = 10.33 + 1.628(2.703) = 14.73$$

$$\text{CL}_{\overline{x}} = \overline{\overline{x}} = 10.33 \qquad \text{(Equations 5-2, 5-27, 5-28)}$$

$$\text{LCL}_{\overline{x}} = \overline{\overline{x}} - A_3\overline{s} = 10.33 - 1.628(2.703) = 5.92$$

$$\text{UCL}_s = B_4\overline{s} = 2.266(2.703) = 6.125$$

$$\text{CL}_s = \overline{s} = 2.703$$

$$\text{LCL}_s = B_3\overline{s} = 0(2.703) = 0$$

MTB>Stat>Control Charts>Variables Charts for Subgroups>Xbar-S (Ex5-2V)

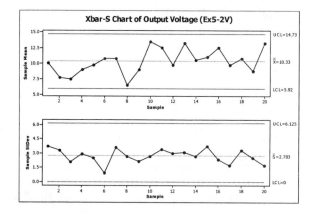

$\overline{x}$ chart: UCL = 14.73, CL = 10.33, LCL = 5.92
s chart: UCL = 6.125, CL = 2.703, LCL = 0

The process is in statistical control with no out-of-control signals, runs, trends, or cycles.

5-9.   Consider the piston ring data shown in Table 5-3. Assume that the specifications on this component are 74.000 ± 0.05 mm.

(a) Set up $\overline{x}$ and R control charts on this process. Is the process in statistical control?

MTB>Stat>Control Charts>Variables Charts for Subgroups>Xbar-R (Ex5-9ID)

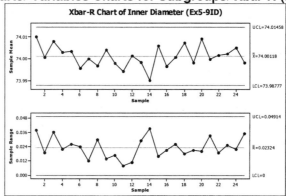

$\overline{x}$ chart: UCL = 74.01458, CL = 74.00118, LCL = 73.98777
R chart: UCL = 0.04914, CL = 0.02324, LCL = 0

The process is in statistical control with no out-of-control signals, runs, trends, or cycles.

(b) Note that the control limits on the $\overline{x}$ chart in part (a) are identical to the control limits on the $\overline{x}$ chart in Example 5-3, where the limits were based on s. Will this always happen?
    The control limits on the $\overline{x}$ charts in Example 5-3 were calculated using $\overline{s}$ to estimate σ, in this exercise $\overline{R}$ was used to estimate σ. They will not always be the same, and in general, the $\overline{x}$ control limits based on $\overline{s}$ will be slightly different than limits based on $\overline{R}$.

5-9 continued

(c) Estimate process capability for the piston-ring process. Estimate the percentage of piston rings produced that will be outside of the specification.

$$\hat{\sigma}_x = \bar{R}/d_2 = 0.02324/2.326 = 0.009991 \text{ and } \hat{C}_p = \frac{USL - LSL}{6\hat{\sigma}_x} = \frac{74.05 - 73.95}{6(0.009991)} = 1.668 \text{ , so the process is}$$

not capable of meeting specifications.

$$\hat{p} = \Pr\{x < LSL\} + \Pr\{x > USL\}$$
$$= \Pr\{x < 73.95\} + \Pr\{x > 74.05\}$$
$$= \Pr\{x < 73.95\} + [1 - \Pr\{x < 74.05\}]$$
$$= \Phi\left(\frac{73.95 - 74.00118}{0.009991}\right) + \left[1 - \Phi\left(\frac{74.05 - 74.00118}{0.009991}\right)\right]$$
$$= \Phi(-5.123) + 1 - \Phi(4.886)$$
$$= 0 + 1 - 1$$
$$= 0$$

MTB>Stat>Quality Tools>Capability Analysis>Normal (Ex5-9ID)
(Under "Estimate" select Rbar as method to estimate standard deviation.)

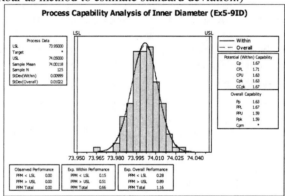

5-11.   Control charts on $\bar{x}$ and s are to be maintained on the torque readings of a bearing used in a wingflap actuator assembly. Samples of size $n = 10$ are to be used, and we know from past experience that when the process is in control, bearing torque has a normal distribution with mean $\mu = 80$ inch-pounds and standard deviation $\sigma = 10$ inch-pounds. Find the center line and control limits for these control charts.

$n = 10$; $\mu = 80$ in-lb; $\sigma_x = 10$ in-lb; and $A = 0.949$; $B_6 = 1.669$; $B_5 = 0.276$ (Appendix VI)

centerline$_{\bar{x}} = \mu = 80$

$UCL_{\bar{x}} = \mu + A\sigma_x = 80 + 0.949(10) = 89.49$

$LCL_{\bar{x}} = \mu - A\sigma_x = 80 - 0.949(10) = 70.51$

centerline$_S = c_4\sigma_x = 0.9727(10) = 9.727$ (Equations 5-15 and 5-25)

$UCL_S = B_6\sigma_x = 1.669(10) = 16.69$

$LCL_S = B_5\sigma_x = 0.276(10) = 2.76$

5-13. Samples of $n = 4$ items are taken from a manufacturing process at regular intervals. A normally distributed quality characteristic is measured and $\bar{x}$ and s values are calculated for each sample. After 50 subgroups have been analyzed, we have $\sum_{i=1}^{50} \bar{x}_i = 1000$ and $\sum_{i=1}^{50} s_i = 72$.

(a) Compute the control limit for the $\bar{x}$ and s control charts.

$$n = 4 \text{ items/subgroup}; \sum_{i=1}^{50} \bar{x}_i = 1000; \sum_{i=1}^{50} S_i = 72; m = 50 \text{ subgroups}$$

$$\bar{\bar{x}} = \frac{\sum_{i=1}^{50} \bar{x}_i}{m} = \frac{1000}{50} = 20; \quad \bar{S} = \frac{\sum_{i=1}^{50} S_i}{m} = \frac{72}{50} = 1.44$$

$$UCL_{\bar{x}} = \bar{\bar{x}} + A_3 \bar{S} = 20 + 1.628(1.44) = 22.34$$

$$LCL_{\bar{x}} = \bar{\bar{x}} - A_3 \bar{S} = 20 - 1.628(1.44) = 17.66 \qquad \text{(Equations 5-27 and 5-28)}$$

$$UCL_S = B_4 \bar{S} = 2.266(1.44) = 3.26$$

$$LCL_S = B_3 \bar{S} = 0(1.44) = 0$$

(b) Assume that all points on both charts plot within the control limits. What are the natural tolerance limits of the process?

Natural process tolerance limits: $\bar{\bar{x}} \pm 3\hat{\sigma}_x = \bar{\bar{x}} \pm 3(\bar{S}/c_4) = 20 \pm 3(1.44/0.9213) = [15.3, 24.7]$

(c) If the specification limits are 19 ± 4.0, what are your conclusions regarding the ability of the process to produce items conforming to specifications?

$$\hat{C}_P = \frac{USL - LSL}{6\hat{\sigma}_x} = \frac{+4.0 - (-4.0)}{6(1.44/0.9213)} = 0.85 \text{, so the process is not capable.}$$

(d) Assuming that if an item exceeds the upper specification limit it can be reworked, and if it is below the lower specification limit it must be scrapped, what percent scrap and rework is the process now producing?

$$\hat{p}_{\text{rework}} = \Pr\{x > USL\} = 1 - \Pr\{x \le USL\} = 1 - \Phi\left(\frac{23 - 20}{1.44/0.9213}\right) = 1 - \Phi(1.919) = 1 - 0.9725 = 0.0275,$$

or 2.75%.

$$\hat{p}_{\text{scrap}} = \Pr\{x < LSL\} = \Phi\left(\frac{15 - 20}{1.44/0.9213}\right) = \Phi(-3.199) = 0.00069 \text{, or } 0.069\%$$

Total $= 2.88\% + 0.069\% = 2.949\%$

(e) If the process were centered at $\mu = 19.0$, what would be the effect on percent scrap and rework?

$$\hat{p}_{\text{rework}} = 1 - \Phi\left(\frac{23 - 19}{1.44/0.9213}\right) = 1 - \Phi(2.56) = 1 - 0.99477 = 0.00523 \text{, or } 0.523\%$$

$$\hat{p}_{\text{scrap}} = \Phi\left(\frac{15 - 19}{1.44/0.9213}\right) = \Phi(-2.56) = 0.00523 \text{, or } 0.523\%$$

Total $= 0.523\% + 0.523\% = 1.046\%$

Centering the process would reduce rework, but increase scrap. A cost analysis is needed to make the final decision. An alternative would be to work to improve the process by reducing variability.

5-15. Parts manufactured by an injection molding process are subjected to a compressive strength test. Twenty samples of five parts each are collected, and the compressive strengths (in psi) are shown in the following table.

Sample Number	x_1	x_2	x_3	x_4	x_5	$\bar{x}$	R
1	83.0	81.2	78.7	75.7	77.0	79.1	7.3
2	88.6	78.3	78.8	71.0	84.2	80.2	17.6
3	85.7	75.8	84.3	75.2	81.0	80.4	10.4
4	80.8	74.4	82.5	74.1	75.7	77.5	8.4
5	83.4	78.4	82.6	78.2	78.9	80.3	5.2
6	75.3	79.9	87.3	80.7	81.8	82.8	14.5
7	74.5	78.0	80.8	73.4	79.7	77.3	7.4
8	79.2	84.4	81.5	86.0	74.5	81.1	11.4
9	80.5	86.2	76.2	64.1	80.2	81.4	9.9
10	75.7	75.2	71.1	82.1	74.3	75.7	10.9
11	80.0	81.5	78.4	73.8	78.1	78.4	7.7
12	80.6	81.8	79.3	73.8	81.7	79.4	8.0
13	82.7	81.3	79.1	82.0	79.5	80.9	3.6
14	79.2	74.9	78.6	77.7	75.3	77.1	4.3
15	85.5	82.1	82.8	73.4	71.7	79.1	13.8
16	78.8	79.6	80.2	79.1	80.8	79.7	2.0
17	82.1	78.2	75.5	78.2	82.1	79.2	6.6
18	84.5	76.9	83.5	81.2	79.2	81.1	7.6
19	79.0	77.8	81.2	84.4	81.6	80.8	6.6
20	84.5	73.1	78.6	78.7	80.6	79.1	11.4

(a) Establish $\bar{x}$ and R control charts for compressive strength using these data. Is the process in statistical control?

MTB>Stat>Control Charts>Variables Charts for Subgroups>Xbar-R (Ex5-15aSt)

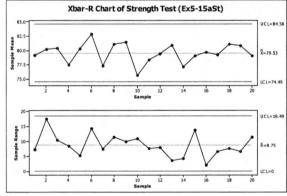

$\bar{x}$ chart: UCL = 84.58, CL = 79.3, LCL = 74.49
R chart: UCL = 18.49, CL = 8.75, LCL = 0

Yes, the process is in control—though we should watch for a possible cyclic pattern in the averages.

5-15 continued

(b) After establishing the control charts in part (a), 15 new subgroups were collected and the compressive strengths are shown next. Plot the $\bar{x}$ and R values against the control units from part (a) and draw conclusions.

MTB>Stat>Control Charts>Variables Charts for Subgroups>Xbar-R

Under "Options, Estimate" use subgroups 1:20 to calculate control limits.

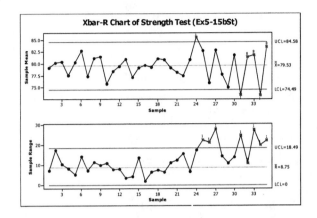

Test Results for R Chart of Ex5-15bSt
```
TEST 1. One point more than 3.00 standard deviations from center line.
Test Failed at points:  25, 26, 27, 31, 33, 34, 35
TEST 2. 9 points in a row on same side of center line.
Test Failed at points:  32, 33, 34, 35
```

A strongly cyclic pattern in the averages is now evident, but more importantly, there are several out-of-control points on the range chart.

5-17. Consider the $\bar{x}$ and R charts you established in Exercise 5-1 using $n = 5$.

(a) Suppose that you wished to continue charting this quality characteristic using $\bar{x}$ and R charts based on a sample size of $n = 3$. What limits would be used on the $\bar{x}$ and R charts?

$n_{old} = 5$; $\bar{\bar{x}}_{old} = 34.00$; $\bar{R}_{old} = 4.7$

For $n_{new} = 3$

$$UCL_{\bar{x}} = \bar{\bar{x}}_{old} + A_{2(new)} \left[\frac{d_{2(new)}}{d_{2(old)}} \right] \bar{R}_{old} = 34 + 1.023 \left[\frac{1.693}{2.326} \right] (4.7) = 37.50$$

$$LCL_{\bar{x}} = \bar{\bar{x}}_{old} - A_{2(new)} \left[\frac{d_{2(new)}}{d_{2(old)}} \right] \bar{R}_{old} = 34 - 1.023 \left[\frac{1.693}{2.326} \right] (4.7) = 30.50$$

$$UCL_R = D_{4(new)} \left[\frac{d_{2(new)}}{d_{2(old)}} \right] \bar{R}_{old} = 2.574 \left[\frac{1.693}{2.326} \right] (4.7) = 8.81$$

$$CL_R = \bar{R}_{new} = \left[\frac{d_{2(new)}}{d_{2(old)}} \right] \bar{R}_{old} = \left[\frac{1.693}{2.326} \right] (4.7) = 3.42$$

$$LCL_R = D_{3(new)} \left[\frac{d_{2(new)}}{d_{2(old)}} \right] \bar{R}_{old} = 0 \left[\frac{1.693}{2.326} \right] (4.7) = 0$$

(b) What would be the impact of the decision you made in part (a) on the ability of the $\bar{x}$ chart to detect a 2σ shift in the mean?

> The $\bar{x}$ control limits for $n = 5$ are "tighter" (31.29, 36.72) than those for $n = 3$ (30.50, 37.50). This means a 2σ shift in the mean would be detected more quickly with a sample size of $n = 5$.

(c) Suppose you wished to continue charting this quality characteristic using $\bar{x}$ and R charts based on a sample size of $n = 8$. What limits would be used on the $\bar{x}$ and R charts?

For $n = 8$

$$UCL_{\bar{x}} = \bar{\bar{x}}_{old} + A_{2(new)} \left[\frac{d_{2(new)}}{d_{2(old)}} \right] \bar{R}_{old} = 34 + 0.373 \left[\frac{2.847}{2.326} \right] (4.7) = 36.15$$

$$LCL_{\bar{x}} = \bar{\bar{x}}_{old} - A_{2(new)} \left[\frac{d_{2(new)}}{d_{2(old)}} \right] \bar{R}_{old} = 34 - 0.373 \left[\frac{2.847}{2.326} \right] (4.7) = 31.85$$

$$UCL_R = D_{4(new)} \left[\frac{d_{2(new)}}{d_{2(old)}} \right] \bar{R}_{old} = 1.864 \left[\frac{2.847}{2.326} \right] (4.7) = 10.72$$

$$CL_R = \bar{R}_{new} = \left[\frac{d_{2(new)}}{d_{2(old)}} \right] \bar{R}_{old} = \left[\frac{2.847}{2.326} \right] (4.7) = 5.75$$

$$LCL_R = D_{3(new)} \left[\frac{d_{2(new)}}{d_{2(old)}} \right] \bar{R}_{old} = 0.136 \left[\frac{2.847}{2.326} \right] (4.7) = 0.78$$

(d) What is the impact of using $n = 8$ on the ability of the $\bar{x}$ chart to detect a 2σ shift in the mean?

> The $\bar{x}$ control limits for $n = 8$ are even "tighter" (31.85, 36.15), increasing the ability of the chart to quickly detect the 2σ shift in process mean.

5-19. Control charts for $\bar{x}$ and R are maintained for an important quality characteristic. The sample size is $n = 7$; $\bar{x}$ and R are computed for each sample. After 35 samples, we have found that $\sum_{i=1}^{35}\bar{x}_i = 7805$ and $\sum_{i=1}^{35}R_i = 1200$.

(a) Set up $\bar{x}$ and R charts using these data.

$$n = 7; \quad \sum_{i=1}^{35}\bar{x}_i = 7805; \quad \sum_{i=1}^{35}R_i = 1200; \quad m = 35 \text{ samples}$$

$$\bar{\bar{x}} = \sum_{i=1}^{35}\bar{x}_i \Big/ m = 7805/35 = 223; \quad \bar{R} = \sum_{i=1}^{35}R_i \Big/ m = 1200/35 = 34.29$$

$$\text{UCL}_{\bar{x}} = \bar{\bar{x}} + A_2\bar{R} = 223 + 0.419(34.29) = 237.37$$

$$\text{LCL}_{\bar{x}} = \bar{\bar{x}} - A_2\bar{R} = 223 - 0.419(34.29) = 208.63 \qquad \text{(Equations 5-4 and 5-5)}$$

$$\text{UCL}_R = D_4\bar{R} = 1.924(34.29) = 65.97$$

$$\text{LCL}_R = D_3\bar{R} = 0.076(34.29) = 2.61$$

(b) Assuming that both charts exhibit control, estimate the process mean and standard deviation.

$$\hat{\mu} = \bar{\bar{x}} = 223; \quad \hat{\sigma}_x = \bar{R}/d_2 = 34.29/2.704 = 12.68 \quad \text{(Equation 5-6)}$$

(c) If the quality characteristic is normally distributed and if the specifications are 220 ± 35, can the process meet the specifications? Estimate the fraction nonconforming.

$$\hat{C}_P = \frac{\text{USL} - \text{LSL}}{6\hat{\sigma}_x} = \frac{+35 - (-35)}{6(12.68)} = 0.92 \text{, the process is not capable of meeting specifications. (Eqn 5-11)}$$

$$\hat{p} = \Pr\{x > \text{USL}\} + \Pr\{x < \text{LSL}\} = 1 - \Pr\{x < \text{USL}\} + \Pr\{x < \text{LSL}\} = 1 - \Pr\{x \le 255\} + \Pr\{x \le 185\}$$

$$= 1 - \Phi\left(\frac{255 - 223}{12.68}\right) + \Phi\left(\frac{185 - 223}{12.68}\right) = 1 - \Phi(2.52) + \Phi(-3.00) = 1 - 0.99413 + 0.00135 = 0.0072 \qquad \text{(p. 202)}$$

(d) Assuming the variance to remain constant, state where the process mean should be located to minimize the fraction nonconforming. What would be the value of the fraction nonconforming under these conditions?

The process mean should be located at the nominal dimension, 220, to minimize non-conforming units.

$$\hat{p} = 1 - \Phi\left(\frac{255 - 220}{12.68}\right) + \Phi\left(\frac{185 - 220}{12.68}\right) = 1 - \Phi(2.76) + \Phi(-2.76) = 1 - 0.99711 + 0.00289 = 0.00578$$

5-21. Samples of size $n = 5$ are collected from a process every half hour. After 50 samples have been collected, we calculate $\bar{x} = 20.0$ and $\bar{s} = 1.5$. Assume that both charts exhibit control and that the quality characteristic is normally distributed.

(a) Estimate the process standard deviation.

$$n = 5; \quad \bar{\bar{x}} = 20.0; \quad \bar{S} = 1.5; \quad m = 50 \text{ samples}$$

$$\hat{\sigma}_x = \bar{s}/c_4 = 1.5/0.9400 = 1.60$$

5-21 continued

(b) Find the control limits on the $\bar{x}$ and s charts.

$$\text{UCL}_{\bar{x}} = \bar{\bar{x}} + A_3\bar{S} = 20.0 + 1.427(1.5) = 22.14$$

$$\text{LCL}_{\bar{x}} = \bar{\bar{x}} - A_3\bar{S} = 20.0 - 1.427(1.5) = 17.86$$

$$\text{UCL}_S = B_4\bar{S} = 2.089(1.5) = 3.13$$

$$\text{LCL}_S = B_3\bar{S} = 0(1.5) = 0$$

(c) If the process mean shifts to 22, what is the probability of concluding that the process is still in control?

$$\text{Pr}\{\text{in control}\} = \text{Pr}\{\text{LCL} \leq \bar{x} \leq \text{UCL}\} = \text{Pr}\{\bar{x} \leq \text{UCL}\} - \text{Pr}\{\bar{x} \leq \text{LCL}\}$$

$$= \Phi\left(\frac{22.14 - 22}{1.6/\sqrt{5}}\right) - \Phi\left(\frac{17.86 - 22}{1.6/\sqrt{5}}\right) = \Phi(0.20) - \Phi(-5.79)$$

$$= 0.57926 - 0 = 0.57926$$

5-23. A critical dimension of a machined part has specifications 100 ± 10. Control chart analysis indicates that the process is in control with $\bar{x} = 104$ and $\bar{R} = 930$. The control charts use samples of size $n = 5$. If we assume that the characteristic is normally distributed, can the mean be located (by adjusting the tool position) so that all output meets specifications? What is the present capability of the process?

$$X \sim N; \quad n = 5; \quad \bar{x} = 104; \quad \bar{R} = 9.30; \quad \text{USL} = 110; \quad \text{LSL} = 90$$

$\hat{\sigma}_x = \bar{R}/d_2 = 9.30/2.326 = 3.998$ and $6\hat{\sigma}_x = 6(3.998) = 23.99$ is larger than the width of the tolerance band, $2(10) = 20$. So, even if the mean is located at the nominal dimension, 100, not all of the output will meet specification.

$$\hat{C}_P = \frac{\text{USL} - \text{LSL}}{6\hat{\sigma}_x} = \frac{+10 - (-10)}{6(3.998)} = 0.8338$$

5-25. Samples of $n = 5$ units are taken from a process every hour. The $\bar{x}$ and R values for a particular quality characteristic are determined. After 25 samples have been collected, we calculate $\bar{x} = 20$ and $\bar{R} = 4.56$.

(a) What are the three-sigma control limits for $\bar{x}$ and R?

$$n = 5; \quad \bar{\bar{x}} = 20; \quad \bar{R} = 4.56; \quad m = 25 \text{ samples}$$

$$\text{UCL}_{\bar{x}} = \bar{\bar{x}} + A_2\bar{R} = 20 + 0.577(4.56) = 22.63$$

$$\text{LCL}_{\bar{x}} = \bar{\bar{x}} - A_2\bar{R} = 20 - 0.577(4.56) = 17.37$$

$$\text{UCL}_R = D_4\bar{R} = 2.114(4.56) = 9.64$$

$$\text{LCL}_R = D_3\bar{R} = 0(4.56) = 0$$

(b) Both charts exhibit control. Estimate the process standard deviation.

$$\hat{\sigma}_x = \bar{R}/d_2 = 4.56/2.326 = 1.96$$

5-25 continued

(c) Assume that the process output is normally distributed. If the specifications are 19 ± 5, what are your conclusions regarding the process capability?

$$\hat{C}_P = \frac{\text{USL} - \text{LSL}}{6\hat{\sigma}_x} = \frac{+5 - (-5)}{6(1.96)} = 0.85 \text{ , so the process is not capable of meeting specifications.}$$

(d) If the process mean shifts to 24, what is the probability of not detecting this shift on the first subsequent sample?

$$\text{Pr\{not detect\}} = \text{Pr\{LCL} \leq \overline{x} \leq \text{UCL\}} = \text{Pr\{}\overline{x} \leq \text{UCL\}} - \text{Pr\{}\overline{x} \leq \text{LCL\}}$$

$$= \Phi\left(\frac{\text{UCL}_{\overline{x}} - \mu_{new}}{\hat{\sigma}_x / \sqrt{n}}\right) - \Phi\left(\frac{\text{LCL}_{\overline{x}} - \mu_{new}}{\hat{\sigma}_x / \sqrt{n}}\right) = \Phi\left(\frac{22.63 - 24}{1.96/\sqrt{5}}\right) - \Phi\left(\frac{17.37 - 24}{1.96/\sqrt{5}}\right)$$

$$= \Phi(-1.56) + \Phi(-7.56) = 0.05938 - 0 = 0.05938$$

5-27. **Continuation of Exercise 5-26.** The following table contains 10 new subgroups of thickness data. Plot this data on the control charts constructed in Exercise 5-26 (a). Is the process in statistical control?

Subgroup	x_1	x_2	x_3	x_4
21	454	449	443	461
22	449	441	444	455
23	442	442	442	450
24	443	452	438	430
25	446	459	457	457
26	454	448	445	462
27	458	449	453	438
28	450	449	445	451
29	443	440	443	451
30	457	450	452	437

From Exercise 5-26 (a):

$\overline{x}$ chart: UCL = 461.88, CL = 449.68, LCL = 437.49

R chart: UCL = 38.18, CL = 16.74, LCL = 0

MTB>Stat>Control Charts>Variables Charts for Subgroups>Xbar-R

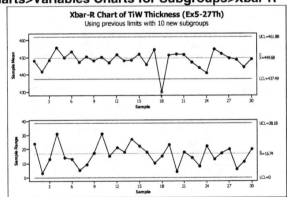

Test Results for Xbar Chart of Ex5-27Th

```
TEST 1. One point more than 3.00 standard deviations from center line.
Test Failed at points:  18
```

The process continues to be in a state of statistical control.

5-29.   Rework Exercises 5-26 and 5-27 using $\bar{x}$ and s control charts.

The process is out of control on the $\bar{x}$ chart at subgroup 18. After finding assignable cause, exclude subgroup 18 from control limits calculations:

MTB>Stat>Control Charts>Variables Charts for Subgroups>Xbar-S

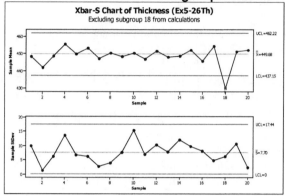

Xbar-S Chart of Ex5-26Th
Test Results for Xbar Chart of Ex5-26Th
```
TEST 1. One point more than 3.00 standard deviations from center line.
Test Failed at points:   18
```

$\bar{x}$ chart: UCL = 462.22, CL = 449.68, LCL = 437.15
s chart: UCL = 17.44, CL = 7.70, LCL = 0

No additional subgroups are beyond the control limits, so these limits can be used for future production.

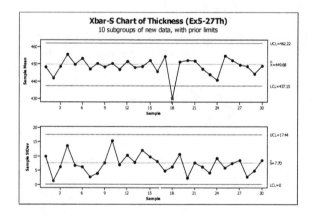

The process remains in statistical control.

5-31. An $\bar{x}$ chart has a center line of 100, uses three-sigma control limits, and is based on a sample size of four. The process standard deviation is known to be six. If the process mean shifts from 100 to 92, what is the probability of detecting this shift on the first sample following the shift?

$$\mu_0 = 100; \quad L = 3; \quad n = 4; \quad \sigma = 6; \quad \mu_1 = 92$$

$$k = (\mu_1 - \mu_0)/\sigma = (92 - 100)/6 = -1.33$$

$$\text{Pr\{detecting shift on 1st sample\}} = 1 - \text{Pr\{not detecting shift on 1st sample\}}$$

$$= 1 - \beta$$

$$= 1 - \left[\Phi\left(L - k\sqrt{n}\right) - \Phi\left(-L - k\sqrt{n}\right) \right]$$

$$= 1 - \left[\Phi\left(3 - (-1.33)\sqrt{4}\right) - \Phi\left(-3 - (-1.33)\sqrt{4}\right) \right] \quad \text{(Equation 5-19)}$$

$$= 1 - \left[\Phi(5.66) - \Phi(-0.34) \right]$$

$$= 1 - \left[1 - 0.37 \right]$$

$$= 0.37$$

5-33. Control charts on $\bar{x}$ and R for samples of size $n = 5$ are to be maintained on the tensile strength in pounds of a yarn. To start the charts, 30 samples were selected, and the mean and range of each computed. This yields $\sum_{i=1}^{30} \bar{x}_i = 607.8$ and $\sum_{i=1}^{30} R_i = 144$.

(a) Compute the center line and control limits for the $\bar{x}$ and R control charts.

$$\bar{\bar{x}} = \sum_{i=1}^{m} \bar{x}_i \bigg/ m = 607.8/30 = 20.26; \quad \bar{R} = \sum_{i=1}^{m} R_i \bigg/ m = 144/30 = 4.8$$

$$\text{UCL}_{\bar{x}} = \bar{\bar{x}} + A_2\bar{R} = 20.26 + 0.577(4.8) = 23.03$$

$$\text{LCL}_{\bar{x}} = \bar{\bar{x}} - A_2\bar{R} = 20.26 - 0.577(4.8) = 17.49 \qquad \text{(Equations 5-4 and 5-5)}$$

$$\text{UCL}_R = D_4\bar{R} = 2.114(4.8) = 10.147$$

$$\text{LCL}_R = D_3\bar{R} = 0(4.8) = 0$$

(b) Suppose both charts exhibit control. There is a single lower specification limit of 16 lb. If strength is normally distributed, what fraction of yarn would fail to meet specifications?

$$\hat{\sigma}_x = \bar{R}/d_2 = 4.8/2.326 = 2.064 \quad \text{(Equation 5-5-6)}$$

$$\hat{p} = \text{Pr}\{x < \text{LSL}\} = \Phi\left(\frac{16 - 20.26}{2.064}\right) = \Phi(-2.064) = 0.0195 \quad \text{(from Appendix II)}$$

5-35. **Continuation of Exercise 5-34.** Reconsider the data from Exercise 5-34 and establish $\bar{x}$ and R charts with appropriate trial control limits. Revise these trial limits as necessary to produce a set of control charts for monitoring future production. Suppose that the following new data are observed.

Sample Number	x_1	x_2	x_3	x_4	x_5
16	2	10	9	6	5
17	1	9	5	9	4
18	0	9	8	2	5
19	−3	0	5	1	4
20	2	10	9	3	1
21	−5	4	0	6	−1
22	0	2	−5	4	6
23	10	0	3	1	5
24	−1	2	5	6	−3
25	0	−1	2	5	−2

MTB>Stat>Control Charts>Variables Charts for Subgroups>R
Under "Options, Estimate" use subgroups 1:11 and 13:15, and select Rbar.

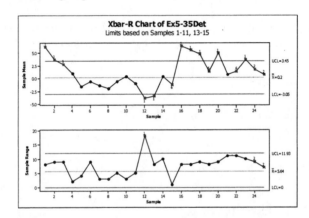

Test Results for Xbar Chart of Ex5-35Det
```
TEST 1. One point more than 3.00 standard deviations from center line.
Test Failed at points:  1, 2, 12, 13, 16, 17, 18, 20, 23
TEST 2. 9 points in a row on same side of center line.
Test Failed at points:  24, 25
TEST 5. 2 out of 3 points more than 2 standard deviations from center line (on
     one side of CL).
Test Failed at points:  2, 3, 13, 17, 18, 20
TEST 6. 4 out of 5 points more than 1 standard deviation from center line (on
     one side of CL).
Test Failed at points:  15, 19, 20, 22, 23, 24
```

Test Results for R Chart of Ex5-35Det
```
TEST 1. One point more than 3.00 standard deviations from center line.
Test Failed at points:  12
TEST 2. 9 points in a row on same side of center line.
Test Failed at points:  24, 25
```

5-35 continued

We are trying to establish trial control limits from the first 15 samples to monitor future production. Note that samples 1, 2, 12, and 13 are out of control on the $\overline{x}$ chart. If these samples are removed and the limits recalculated, sample 3 is also out of control on the $\overline{x}$ chart. Removing sample 3 gives

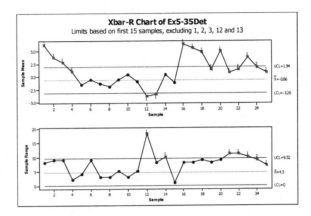

Sample 14 is now out of control on the R chart. No additional samples are out of control on the $\overline{x}$ chart. While the limits on the above charts may be used to monitor future production, the fact that 6 of 15 samples were out of control and eliminated from calculations is an early indication of process instability.

(a) Plot these new observations on the control chart. What conclusions can you draw about process stability?
Given the large number of points after sample 15 beyond both the $\overline{x}$ and R control limits on the charts above, the process appears to be unstable.

(b) Use all 25 observations to revise the control limits for the $\overline{x}$ and R charts. What conclusions can you draw now about the process?
With Test 1 only:

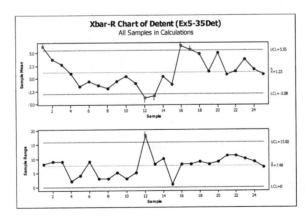

> **Test Results for Xbar Chart of Ex5-35Det**
> TEST 1. One point more than 3.00 standard deviations from center line.
> Test Failed at points: 1, 12, 13, 16, 17
> **Test Results for R Chart of Ex5-35Det**
> TEST 1. One point more than 3.00 standard deviations from center line.
> Test Failed at points: 12

5-35 continued

Removing samples 1, 12, 13, 16, and 17 from calculations, and with Test 1 only:

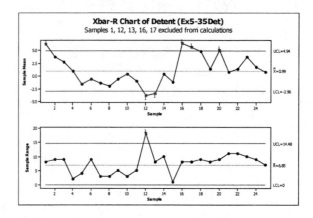

> **Test Results for Xbar Chart of Ex5-35Det**
> TEST 1. One point more than 3.00 standard deviations from center line.
> Test Failed at points: 1, 12, 13, 16, 17, 20
> **Test Results for R Chart of Ex5-35Det**
> TEST 1. One point more than 3.00 standard deviations from center line.
> Test Failed at points: 12

Sample 20 is now also out of control. Removing sample 20 from calculations, and with Test 1 only:

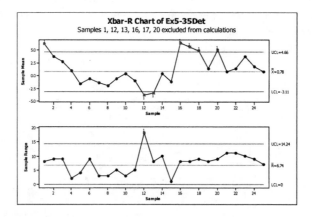

> **Test Results for Xbar Chart of Ex5-35Det**
> TEST 1. One point more than 3.00 standard deviations from center line.
> Test Failed at points: 1, 12, 13, 16, 17, 18, 20
> **Test Results for R Chart of Ex5-35Det**
> TEST 1. One point more than 3.00 standard deviations from center line.
> Test Failed at points: 12

Sample 18 is now out-of-control, for a total 7 of the 25 samples, with runs of points both above and below the centerline. This suggests that the process is inherently unstable, and that the sources of variation need to be identified and removed.

5-37. Control charts for $\bar{x}$ and R are maintained on the tensile strength of a metal fastener. After 30 samples of size $n = 6$ are analyzed, we find that $\sum_{i=1}^{35} \bar{x}_i = 12,870$ and $\sum_{i=1}^{35} R_i = 1350$.

(a) Compute control limits on the R chart.

$$n = 6; \quad \sum_{i=1}^{30} \bar{x}_i = 12,870; \quad \sum_{i=1}^{30} R_i = 1350; \quad m = 30$$

$$\bar{R} = \frac{\sum_{i=1}^{m} R_i}{m} = \frac{1350}{30} = 45.0$$

$$\text{UCL}_R = D_4 \bar{R} = 2.004(45.0) = 90.18$$

$$\text{LCL}_R = D_3 \bar{R} = 0(45.0) = 0$$

(b) Assuming that the R chart exhibits control, estimate the parameters μ and σ.

$$\hat{\mu} = \bar{\bar{x}} = \frac{\sum_{i=1}^{m} \bar{x}_i}{m} = \frac{12,870}{30} = 429.0$$

$$\hat{\sigma}_x = \bar{R} / d_2 = 45.0 / 2.534 = 17.758$$

(c) If the process output is normally distributed, and if the specifications are 440 ± 40, can the process meet the specifications? Estimate the fraction nonconforming.

$$\text{USL} = 440 + 40 = 480; \quad \text{LSL} = 440 - 40 = 400$$

$$\hat{C}_p \frac{\text{USL} - \text{LSL}}{6\hat{\sigma}_x} = \frac{480 - 400}{6(17.758)} = 0.751$$

$$\hat{p} = 1 - \Phi\left(\frac{480 - 429}{17.758}\right) + \Phi\left(\frac{400 - 429}{17.758}\right) = 1 - \Phi(2.87) + \Phi(-1.63) = 1 - 0.9979 + 0.0516 = 0.0537$$

(d) If the variance remains constant, where should the mean be located to minimize the fraction nonconforming?

To minimize fraction nonconforming the mean should be located at the nominal dimension (440) for a constant variance.

5-39.   An $\bar{x}$ chart on a normally distributed quality characteristic is to be established with the standard values $\mu = 100$, $\sigma = 8$, and $n = 4$. Find the following:

(a) The two-sigma control limits.

$$n = 4; \quad \mu = 100; \quad \sigma_x = 8$$

$$\text{UCL}_{\bar{x}} = \mu + 2\sigma_{\bar{x}} = \mu + 2\left(\sigma_x / \sqrt{n}\right) = 100 + 2\left(8/\sqrt{4}\right) = 108$$

$$\text{LCL}_{\bar{x}} = \mu - 2\sigma_{\bar{x}} = \mu - 2\left(\sigma_x / \sqrt{n}\right) = 100 - 2\left(8/\sqrt{4}\right) = 92$$

(b) The 0.005 probability limits.

$$k = Z_{\alpha/2} = Z_{0.005/2} = Z_{0.0025} = 2.807$$

$$\text{UCL}_{\bar{x}} = \mu + k\sigma_{\bar{x}} = \mu + k\left(\sigma_x / \sqrt{n}\right) = 100 + 2.807\left(8/\sqrt{4}\right) = 111.228$$

$$\text{LCL}_{\bar{x}} = \mu - k\sigma_{\bar{x}} = \mu - k\left(\sigma_x / \sqrt{n}\right) = 100 - 2.807\left(8/\sqrt{4}\right) = 88.772$$

5-41.   Consider the $\bar{x}$ chart defined in Exercise 5-40. Find the ARL_1 for the chart.

From Exercise 5-40:

$$\beta = \text{Pr\{not detected\}} = \text{Pr\{LCL}_{\bar{x}} \le \bar{x} \le \text{UCL}_{\bar{x}}\} = \text{Pr\{}\bar{x} \le \text{UCL}_{\bar{x}}\} - \text{Pr\{}\bar{x} \le \text{LCL}_{\bar{x}}\}$$

$$= \Phi\left(\frac{\text{UCL}_{\bar{x}} - \mu}{\sigma_{\bar{x}}}\right) - \Phi\left(\frac{\text{LCL}_{\bar{x}} - \mu}{\sigma_{\bar{x}}}\right) = \Phi\left(\frac{104 - 98}{8/\sqrt{5}}\right) - \Phi\left(\frac{96 - 98}{8/\sqrt{5}}\right) = \Phi(1.68) - \Phi(-0.56)$$

$$= 0.9535 - 0.2877 = 0.6658$$

$$\text{ARL}_1 = \frac{1}{1 - \beta} = \frac{1}{1 - \text{Pr\{not detect\}}} = \frac{1}{1 - 0.6658} = 2.992$$

5-43.   Statistical monitoring of a quality characteristic uses both an $\bar{x}$ and an s chart. The charts are to be based on the standard values $\mu = 200$ and $\sigma = 10$, with $n = 4$.

(a) Find three-sigma control limits for the s chart.

$$\text{centerline}_s = c_4\sigma = 0.9213(10) = 9.213$$

$$\text{UCL}_s = B_6\sigma_x = 2.088(10) = 20.88$$

$$\text{LCL}_s = B_5\sigma_x = 0(10) = 0$$

(b) Find a center line and control limits for the $\bar{x}$ chart such that the probability of a type I error is 0.05.

$$k = Z_{\alpha/2} = Z_{0.05/2} = Z_{0.025} = 1.96$$

$$\text{UCL}_{\bar{x}} = \mu + k\sigma_{\bar{x}} = \mu + k\left(\sigma_x / \sqrt{n}\right) = 200 + 1.96\left(10/\sqrt{4}\right) = 209.8$$

$$\text{LCL}_{\bar{x}} = \mu - k\sigma_{\bar{x}} = \mu - k\left(\sigma_x / \sqrt{n}\right) = 200 - 1.96\left(10/\sqrt{4}\right) = 190.2$$

5-45. Thirty samples each of size seven have been collected to establish control over a process. The following data were collected: $\sum_{i=1}^{30} \bar{x}_i = 2700$ and $\sum_{i=1}^{30} R_i = 120$

(a) Calculate trial control limits for the two charts.

$$n = 7; \quad \sum_{i=1}^{30} \bar{x}_i = 2700; \quad \sum_{i=1}^{30} R_i = 120; \quad m = 30$$

$$\bar{\bar{x}} = \frac{\sum_{i=1}^{m} \bar{x}_i}{m} = \frac{2700}{30} = 90; \quad \bar{R} = \frac{\sum_{i=1}^{m} R_i}{m} = \frac{120}{30} = 4$$

$$\mathrm{UCL}_{\bar{x}} = \bar{\bar{x}} + A_2\bar{R} = 90 + 0.419(4) = 91.676$$

$$\mathrm{LCL}_{\bar{x}} = \bar{\bar{x}} - A_2\bar{R} = 90 - 0.419(4) = 88.324$$

$$\mathrm{UCL}_R = D_4\bar{R} = 1.924(4) = 7.696$$

$$\mathrm{LCL}_R = D_3\bar{R} = 0.076(4) = 0.304$$

(b) On the assumption that the R chart is in control, estimate the process standard deviation.

$$\hat{\sigma}_x = \bar{R}/d_2 = 4/2.704 = 1.479$$

(c) Suppose an s chart were desired. What would be the appropriate control limits and center line?

$$\bar{S} = c_4\hat{\sigma}_x = 0.9594(1.479) = 1.419$$

$$\mathrm{UCL}_S = 1.882(1.419) = 2.671$$

$$\mathrm{LCL}_S = 0.118(1.419) = 0.167$$

5-47. $\bar{x}$ and R charts with $n = 4$ are used to monitor a normally distributed quality characteristic. The control chart parameters are

$\bar{x}$ Chart	R Chart
UCL = 815	UCL = 46.98
Center line = 800	Center line = 20.59
LCL = 785	LCL = 0

Both charts exhibit control. What is the probability that a shift in the process mean to 790 will be detected on the first sample following the shift?

$$\hat{\sigma}_x = \bar{R}/d_2 = 20.59/2.059 = 10$$

$$\Pr\{\text{detect shift on 1st sample}\} = \Pr\{\bar{x} < \mathrm{LCL}\} + \Pr\{\bar{x} > \mathrm{UCL}\} = \Pr\{\bar{x} < \mathrm{LCL}\} + 1 - \Pr\{\bar{x} \le \mathrm{UCL}\}$$

$$= \Phi\left(\frac{\mathrm{LCL} - \mu_{\text{new}}}{\sigma_{\bar{x}}}\right) + 1 - \Phi\left(\frac{\mathrm{UCL} - \mu_{\text{new}}}{\sigma_{\bar{x}}}\right) = \Phi\left(\frac{785 - 790}{10/\sqrt{4}}\right) + 1 - \Phi\left(\frac{815 - 790}{10/\sqrt{4}}\right)$$

$$= \Phi(-1) + 1 - \Phi(5) = 0.1587 + 1 - 1.0000 = 0.1587$$

5-49.   Control charts for $\bar{x}$ and R are in use with the following parameters:

$\bar{x}$ Chart	R Chart
UCL = 363.0	UCL = 16.18
Center line = 360.0	Center line = 8.91
LCL = 357.0	LCL = 1.64

The sample size is $n = 9$. Both charts exhibit control. The quality characteristic is normally distributed.

(a) What is the α–risk associated with the $\bar{x}$ chart?

$$\hat{\sigma}_x = \bar{R}/d_2 = 8.91/2.970 = 3.000$$

$$\alpha = \Pr\{\bar{x} < \text{LCL}\} + \Pr\{\bar{x} > \text{UCL}\} = \Phi\left(\frac{\text{LCL} - \bar{\bar{x}}}{\sigma_{\bar{x}}}\right) + 1 - \Phi\left(\frac{\text{UCL} - \bar{\bar{x}}}{\sigma_{\bar{x}}}\right)$$

$$= \Phi\left(\frac{357 - 360}{3/\sqrt{9}}\right) + 1 - \Phi\left(\frac{363 - 360}{3/\sqrt{9}}\right) = \Phi(-3) + 1 - \Phi(3) = 0.0013 + 1 - 0.9987 = 0.0026$$

(b) Specifications on this quality characteristic are 358 ± 6. What are your conclusions regarding the ability of the process to produce items within specifications?

$$\hat{C}_P = \frac{\text{USL} - \text{LSL}}{6\hat{\sigma}_x} = \frac{+6 - (-6)}{6(3)} = 0.667$$

The process is not capable of producing all items within specification.

(c) Suppose the mean shifts to 357. What is the probability that the shift will not be detected on the first sample following the shift?

$$\mu_{\text{new}} = 357$$

$$\Pr\{\text{not detect on 1st sample}\} = \Pr\{\text{LCL} \leq \bar{x} \leq \text{UCL}\} = \Phi\left(\frac{\text{UCL} - \mu_{\text{new}}}{\hat{\sigma}_x/\sqrt{n}}\right) - \Phi\left(\frac{\text{LCL} - \mu_{\text{new}}}{\hat{\sigma}_x/\sqrt{n}}\right)$$

$$= \Phi\left(\frac{363 - 357}{3/\sqrt{9}}\right) - \Phi\left(\frac{357 - 357}{3/\sqrt{9}}\right) = \Phi(6) - \Phi(0) = 1.0000 - 0.5000 = 0.5000$$

(d) What would be the appropriate control limits for the $\bar{x}$ chart if the type I error probability were to be 0.01? A normally distributed quality characteristic

$$\alpha = 0.01; \quad k = Z_{\alpha/2} = Z_{0.01/2} = Z_{0.005} = 2.576$$

$$\text{UCL}_{\bar{x}} = \bar{\bar{x}} + k\sigma_{\bar{x}} = \bar{\bar{x}} + k\left(\hat{\sigma}_x/\sqrt{n}\right) = 360 + 2.576\left(3/\sqrt{9}\right) = 362.576$$

$$\text{LCL}_{\bar{x}} = 360 - 2.576\left(3/\sqrt{9}\right) = 357.424$$

5-51.   Control charts for $\bar{x}$ and s have been maintained on a process and have exhibited statistical control. The sample size is $n = 6$. The control chart parameters are as follows:

$\bar{x}$ Chart	S Chart
UCL = 708.20	UCL = 3.420
Center line = 706.00	Center line = 1.738
LCL = 703.80	LCL = 0.052

(a) Estimate the mean and standard deviation of the process.

$\hat{\mu} = \bar{\bar{x}} = 706.00; \quad \hat{\sigma}_x = \bar{S}/c_4 = 1.738/0.9515 = 1.827$

(b) Estimate the natural tolerance limits for the process.

$\text{UNTL} = \bar{\bar{x}} + 3\hat{\sigma}_x = 706 + 3(1.827) = 711.48$

$\text{LNTL} = 706 - 3(1.827) = 700.52$

(c) Assume that the process output is well modeled by a normal distribution. If specifications are 703 and 709, estimate the fraction nonconforming.

$\hat{p} = \Pr\{x < \text{LSL}\} + \Pr\{x > \text{USL}\}$

$= \Phi\left(\dfrac{\text{LSL} - \bar{\bar{x}}}{\hat{\sigma}_x}\right) + 1 - \Phi\left(\dfrac{\text{USL} - \bar{\bar{x}}}{\hat{\sigma}_x}\right)$

$= \Phi\left(\dfrac{703 - 706}{1.827}\right) + 1 - \Phi\left(\dfrac{709 - 706}{1.827}\right)$

$= \Phi(-1.642) + 1 - \Phi(1.642) = 0.0503 + 1 - 0.9497 = 0.1006$

(d) Suppose the process mean shifts to 702.00 while the standard deviation remains constant. What is the probability of an out-of-control signal occurring on the first sample following the shift?

$\Pr\{\text{detect on 1st sample}\} = \Pr\{\bar{x} < \text{LCL}\} + \Pr\{\bar{x} > \text{UCL}\}$

$= \Phi\left(\dfrac{\text{LCL} - \mu_{\text{new}}}{\sigma_{\bar{x}}}\right) + 1 - \Phi\left(\dfrac{\text{UCL} - \mu_{\text{new}}}{\sigma_{\bar{x}}}\right)$

$= \Phi\left(\dfrac{703.8 - 702}{1.827/\sqrt{6}}\right) + 1 - \Phi\left(\dfrac{708.2 - 702}{1.827/\sqrt{6}}\right)$

$= \Phi(2.41) + 1 - \Phi(8.31) = 0.9920 + 1 - 1.0000 = 0.9920$

(e) For the shift in part (d), what is the probability of detecting the shift by at least the third subsequent sample?

$\Pr\{\text{detect by 3rd sample}\} = 1 - \Pr\{\text{not detect by 3rd sample}\}$

$= 1 - (\Pr\{\text{not detect}\})^3 = 1 - (1 - 0.9920)^3 = 1.0000$

5-53.   One-pound coffee cans are filled by a machine, sealed, and then weighed automatically. After adjusting for the weight of the can, any package that weighs less than 16 oz is cut out of the conveyor. The weights of 25 successive cans are shown here. Set up a moving-range control chart and a control chart for individuals. Estimate the mean and standard deviation of the amount of coffee packed in each can. Is it reasonable to assume that can weight is normally distributed? If the process remains in control at this level, what percentage of cans will be underfilled?

Can Number	Weight	Can Number	Weight
1	16.11	14	16.12
2	16.08	15	16.10
3	16.12	16	16.08
4	16.10	17	16.13
5	16.10	18	16.15
6	16.11	19	16.12
7	16.12	20	16.10
8	16.09	21	16.08
9	16.12	22	16.07
10	16.10	23	16.11
11	16.09	24	16.13
12	16.07	25	16.10
13	16.13		

MTB>Stat>Control Charts>Variables Charts for Individuals>I-MR

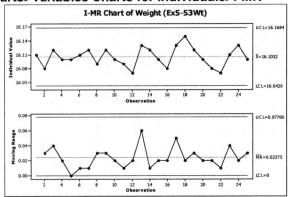

$\bar{x}$ chart: UCL = 16.1684, CL = 16.1052, LCL = 16.0420
R chart: UCL = 0.07760, CL = 0.02375, LCL = 0

There may be a "sawtooth" pattern developing on the Individuals chart.

$\bar{\bar{x}} = 16.1052$; $\hat{\sigma}_x = 0.021055$; $\overline{MR2} = 0.02375$

5-53 continued

MTB>Stat>Basic Statistics>Normality Test

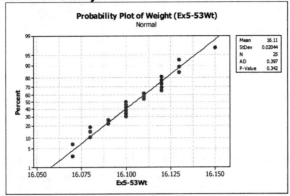

Visual examination of the normal probability indicates that the assumption of normally distributed coffee can weights is valid.

$$\%\text{underfilled} = 100\% \times \Pr\{x < 16 \text{ oz}\}$$

$$= 100\% \times \Phi\left(\frac{16 - 16.1052}{0.021055}\right) = 100\% \times \Phi(-4.9964) = 0.00003\%$$

5-55.   The viscosity of a polymer is measured hourly. Measurements for the last 20 hours are shown as follows:

Test	Viscosity	Test	Viscosity
1	2838	11	3174
2	2785	12	3102
3	3058	13	2762
4	3064	14	2975
5	2996	15	2719
6	2882	16	2861
7	2878	17	2797
8	2920	18	3078
9	3050	19	2964
10	2870	20	2805

(a) Does viscosity follow a normal distribution?

MTB>Stat>Basic Statistics>Normality Test

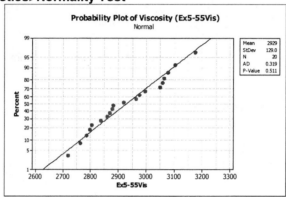

Viscosity measurements do appear to follow a normal distribution.

(b) Set up a control chart on viscosity and a moving range chart. Does the process exhibit statistical control?

MTB>Stat>Control Charts>Variables Charts for Individuals>I-MR

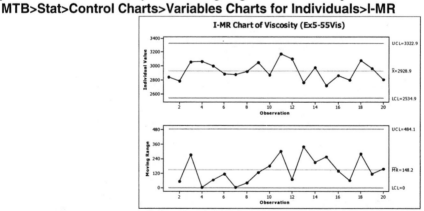

$\bar{x}$ chart: UCL = 3322.9, CL = 2928.9, LCL = 2534.9
R chart: UCL = 484.1, CL = 148.2, LCL = 0

The process appears to be in statistical control, with no out-of-control points, runs, trends, or other patterns.

(c) Estimate the process mean and standard deviation.

$$\hat{\mu} = \bar{\bar{x}} = 2928.9; \quad \hat{\sigma}_x = 131.346; \quad \overline{MR2} = 148.158$$

5-57. (a) Thirty observations on the oxide thickness of individual silicon wafers are shown here. Use these data to set up a control chart on oxide thickness and a moving range chart. Does the process exhibit statistical control? Does oxide thickness follow a normal distribution?

Wafer	Oxide Thickness	Wafer	Oxide Thickness
1	45.4	16	58.4
2	48.6	17	51.0
3	49.5	18	41.2
4	44.0	19	47.1
5	50.9	20	45.7
6	55.2	21	60.6
7	45.5	22	51.0
8	52.8	23	53.0
9	45.3	24	56.0
10	46.3	25	47.2
11	53.9	26	48.0
12	49.8	27	55.9
13	46.9	28	50.0
14	49.8	29	47.9
15	45.1	30	53.4

MTB>Stat>Control Charts>Variables Charts for Individuals>I-MR

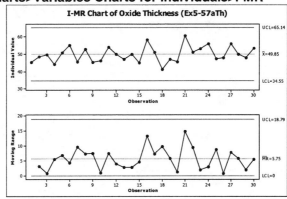

x chart: UCL = 65.14, CL = 49.85, LCL = 34.55
MR chart: UCL = 18.79, CL = 5.75, LCL = 0
The process is in statistical control.

MTB>Stat>Basic Statistics>Normality Test

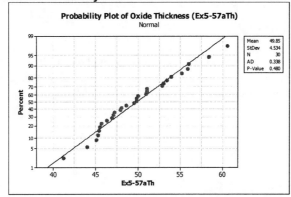

The normality assumption is reasonable.

5-57 continued

(b) Following the establishment of the control charts in part (a), ten new wafers were observed. The oxide thickness measurements are as follows:

Wafer	Oxide Thickness	Wafer	Oxide Thickness
1	54.3	6	51.5
2	57.5	7	58.4
3	64.8	8	67.5
4	62.1	9	61.1
5	59.6	10	63.3

Plot these observations against the control limits determined in part (a). Is the process in control?

MTB>Stat>Control Charts>Variables Charts for Individuals>I-MR

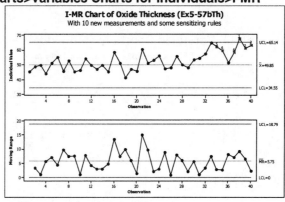

Test Results for I Chart of Ex5-57bTh
```
TEST 1. One point more than 3.00 standard deviations from center line.
Test Failed at points:  38
TEST 2. 9 points in a row on same side of center line.
Test Failed at points:  38, 39, 40
TEST 5. 2 out of 3 points more than 2 standard deviations from center line (on
      one side of CL).
Test Failed at points:  34, 39, 40
TEST 6. 4 out of 5 points more than 1 standard deviation from center line (on
      one side of CL).
Test Failed at points:  35, 37, 38, 39, 40
```

We have turned on some of the sensitizing rules in MINITAB to illustrate their use. There is a run above the centerline, several 4 of 5 beyond 1 sigma, and several 2 of 3 beyond 2 sigma on the *x* chart. However, even without use of the sensitizing rules, it is clear that the process is out of control during this period of operation.

5-57 continued

(c) Suppose the assignable cause responsible for the out-of-control signal in part (b) is discovered and removed from the process. Twenty additional wafers are subsequently sampled. Plot the oxide thickness against the part (a) control limits. What conclusions can you draw? The new data are shown here.

Wafer	Oxide Thickness	Wafer	Oxide Thickness
1	43.4	11	50.0
2	46.7	12	61.2
3	44.8	13	46.9
4	51.3	14	44.9
5	49.2	15	46.2
6	46.5	16	53.3
7	48.4	17	44.1
8	50.1	18	47.4
9	53.7	19	51.3
10	45.6	20	42.5

MTB>Stat>Control Charts>Variables Charts for Individuals>I-MR

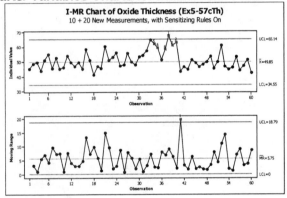

The process has been returned to a state of statistical control.

5-59.   In 1879, A. A. Michelson measured the velocity of light in air using a modification of a method proposed by the French physicist Foucault. Twenty of these measurements follow (the value reported is in kilometers per second and has 299,000 subtracted from it). Use these data to set up an individuals and a moving-range control chart. Is there some evidence that the measurements of the velocity of light are normally distributed? Do the measurements exhibit statistical control? Revise the control limits if necessary.

Measurement	Velocity	Measurement	Velocity
1	850	11	850
2	1000	12	810
3	740	13	950
4	980	14	1000
5	900	15	980
6	930	16	1000
7	1070	17	980
8	650	18	960
9	930	19	880
10	760	20	960

MTB>Stat>Basic Statistics>Normality Test

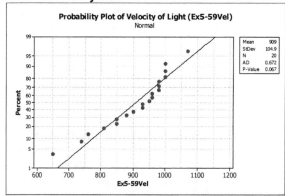

The velocity of light measurements is approximately normally distributed.

5-59 continued

MTB>Stat>Control Charts>Variables Charts for Individuals>I-MR

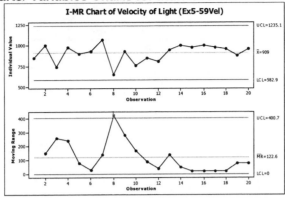

Test Results for MR Chart of Ex5-59Vel
TEST 1. One point more than 3.00 standard deviations from center line.
Test Failed at points: 8

x chart: UCL = 1235.1, CL = 909, LCL = 582.9
MR chart: UCL = 400.7, CL = 122.6, LCL = 0

The out-of-control signal on the moving range chart indicates a significantly large difference between successive measurements (7 and 8). Since neither of these measurements seems unusual, use all data for control limits calculations. There may also be an early indication of less variability in the later measurements. For now, consider the process to be in a state of statistical process control.

5-61.   The uniformity of a silicon wafer following an etching process is determined by measuring the layer thickness at several locations and expressing uniformity as the range of the thicknesses. Following are uniformity determinations for 30 consecutive wafers processed through the etching tool.

Wafer	Uniformity	Wafer	Uniformity
1	11	16	15
2	16	17	16
3	22	18	12
4	14	19	11
5	34	20	18
6	22	21	14
7	13	22	13
8	11	23	18
9	6	24	12
10	11	25	13
11	11	26	12
12	23	27	15
13	14	28	21
14	12	29	21
15	7	30	14

(a) Is there evidence that uniformity is normally distributed? If not, find a suitable transformation for the data.

MTB>Stat>Basic Statistics>Normality Test

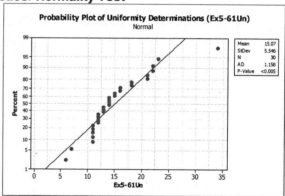

The data are not normally distributed, as evidenced by the "S"- shaped curve to the plot points on a normal probability plot, as well as the Anderson-Darling test p-value.

The data are skewed right, so a compressive transform such as natural log or square-root may be appropriate.

5-61 (a) continued

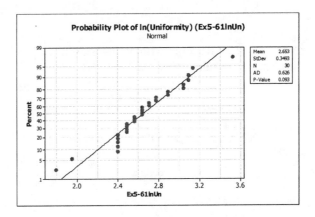

The distribution of the natural-log transformed uniformity measurements is approximately normally distributed.

(b) Construct a control chart for individuals and a moving-range control chart for uniformity for the etching process. Is the process in statistical control?

MTB>Stat>Control Charts>Variables Charts for Individuals>I-MR

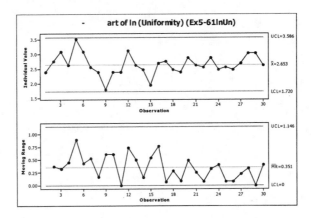

x chart: UCL = 3.586, CL = 2.653, LCL = 1.720
MR chart: UCL = 1.146, CL = 0.351, LCL = 0

The etching process appears to be in statistical control.

5-63. Reconsider the situation in Exercise 5-53. Construct an individuals control chart using the median of the span-two moving ranges to estimate variability. Compare this control chart to the one constructed in Exercise 5-53 and discuss.

MTB>Stat>Control Charts>Variables Charts for Individuals>I-MR
Select "Estimate" to change the method of estimating sigma

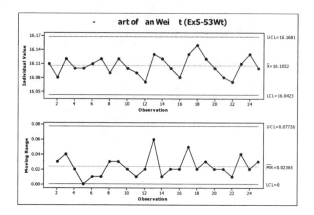

x chart: UCL = 16.1681, CL = 16.1052, LCL = 16.0423
MR chart: UCL = 0.07726, CL = 0.02365, LCL = 0

There is no difference between this chart and the one in Exercise 5-53; control limits for both are essentially the same.

5-65.   Reconsider the polymer viscosity data in Exercise 5-55. Use the median of the span-two moving ranges to estimate σ and set up the individuals control chart. Compare this chart to the one originally constructed using the average moving range method to estimate σ.

MTB>Stat>Control Charts>Variables Charts for Individuals>I-MR
Select "Estimate" to change the method of estimating sigma

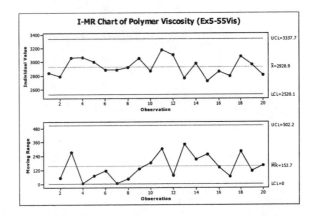

x chart: UCL = 3337.7, CL = 2928.9, LCL = 2520.1
MR chart: UCL = 502.2, CL = 153.7, LCL = 0

The median moving range method gives slightly wider control limits for both the Individual and Moving Range charts, with no practical meaning for this set of observations.

5-67.   Consider the individuals measurement data shown next.

Observation	x	Observation	x
1	10.07	5	8.99
2	10.47	6	7.74
3	9.45	7	10.63
4	9.44	8	9.78

Observation	x	Observation	x
9	9.37	18	11.26
10	9.95	19	9.48
11	12.04	20	11.28
12	10.93	21	12.54
13	11.54	22	11.48
14	9.58	23	13.26
15	8.80	24	11.10
16	12.94	25	10.82
17	10.78		

(a) Estimate σ using the average of the moving ranges of span two.

MTB>Stat>Control Charts>Variables Charts for Individuals>I-MR

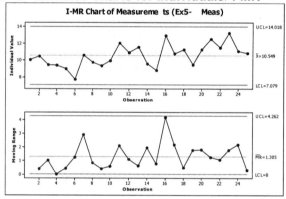

x chart: UCL = 14.018, CL = 105.49, LCL = 7.079
MR chart: UCL = 4.262, CL = 1.305, LCL = 0

$$\hat{\sigma}_x = \overline{R} / d_2 = 1.305 / 1.128 = 1.157$$

(b) Estimate σ using s/c_4.

MTB>Stat>Basic Statistics>Descriptive Statistics

Descriptive Statistics: Ex5-67Meas

Variable	Total Count	Mean	StDev	Median
Ex5-67Meas	25	10.549	1.342	10.630

$$\hat{\sigma}_x = S / c_4 = 1.342 / 0.7979 = 1.682$$

5-67 continued

(c) Estimate σ using the median of the span-two moving ranges.

MTB>Stat>Control Charts>Variables Charts for Individuals>I-MR

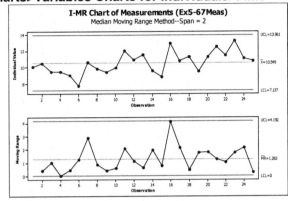

I-MR Chart of Measurements (Ex5-67Meas)
Median Moving Range Method--Span = 2

x chart: UCL = 13.961, CL = 10.549, LCL = 7.137
MR chart: UCL = 4.192, CL = 1.283, LCL = 0

$$\hat{\sigma}_x = \overline{R}/d_2 = 1.283/1.128 = 1.137$$

(d) Estimate σ using the average of the moving ranges of span 3, 4, …, 20.

Average MR3 Chart: $\hat{\sigma}_x = \overline{R}/d_2 = 2.049/1.693 = 1.210$

Average MR4 Chart: $\hat{\sigma}_x = \overline{R}/d_2 = 2.598/2.059 = 1.262$

…

Average MR19 Chart: $\hat{\sigma}_x = \overline{R}/d_2 = 5.186/3.689 = 1.406$

Average MR20 Chart: $\hat{\sigma}_x = \overline{R}/d_2 = 5.36/3.735 = 1.435$

(e) Discuss the results you have obtained.

As the span of the moving range is increased, there are fewer observations to estimate the standard deviation, and the estimate becomes less reliable. For this example, σ gets larger as the span increases. This tends to be true for unstable processes.

5-69. The diameter of the casting in Fig. 5-26 is also an important quality characteristic. A coordinate measuring machine is used to measure the diameter of each casting at five different locations. Data for 20 castings are shown in the following table.

Casting	Diameter				
	1	2	3	4	5
1	11.7629	11.7403	11.7511	11.7474	11.7374
2	11.8122	11.7506	11.7787	11.7736	11.8412
3	11.7742	11.7114	11.7530	11.7532	11.7773
4	11.7833	11.7311	11.7777	11.8108	11.7804
5	11.7134	11.6870	11.7305	11.7419	11.6642
6	11.7925	11.7611	11.7588	11.7012	11.7611
7	11.6916	11.7205	11.6958	11.7440	11.7062
8	11.7109	11.7832	11.7496	11.7496	11.7318
9	11.7984	11.8887	11.7729	11.8485	11.8416
10	11.7914	11.7613	11.7356	11.7628	11.7070
11	11.7260	11.7329	11.7424	11.7645	11.7571
12	11.7202	11.7537	11.7328	11.7582	11.7265
13	11.8356	11.7971	11.8023	11.7802	11.7903
14	11.7069	11.7112	11.7492	11.7329	11.7289
15	11.7116	11.7978	11.7982	11.7429	11.7154
16	11.7165	11.7284	11.7571	11.7597	11.7317
17	11.8022	11.8127	11.7864	11.7917	11.8167
18	11.7775	11.7372	11.7241	11.7773	11.7543
19	11.7753	11.7870	11.7574	11.7620	11.7673
20	11.7572	11.7626	11.7523	11.7395	11.7884

(a) Set up $\bar{x}$ and R charts for this process, assuming the measurements on each casting form a rational subgroup.

MTB>Stat>Control Charts>Variables Charts for Subgroups>Xbar-R

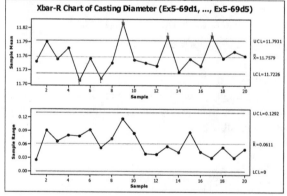

Test Results for Xbar Chart of Ex5-69d1, ..., Ex5-69d5
```
TEST 1. One point more than 3.00 standard deviations from center line.
Test Failed at points:  5, 7, 9, 13, 17
TEST 5. 2 out of 3 points more than 2 standard deviations from center line (on
      one side of CL).
Test Failed at points:  7
```

$\bar{x}$ chart: UCL = 11.7931, CL = 11.7579, LCL = 11.7226
R chart: UCL = 0.1292, CL = 0.0611, LCL = 0

5-69 continued

(b) Discuss the charts you have constructed in part (a).

Though the R chart is in control, plot points on the $\bar{x}$ chart bounce below and above the control limits. Since these are high precision castings, we might expect that the diameter of a single casting will not change much with location. If no assignable cause can be found for these out-of-control points, we may want to consider treating the averages as an Individual value and graphing "between/within" range charts. This will lead to an understanding of the greatest source of variability, between castings or within a casting.

(c) Construct "between/within" charts for this process.

MTB>Stat>Control Charts>Variables Charts for Subgroups>I-MR-R/S (Between/Within)
Select "**I-MR-R/S Options, Estimate**" and choose R-bar method to estimate standard deviation

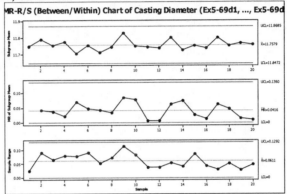

I-MR-R/S Standard Deviations of Ex5-69d1, ..., Ex5-69d5

Standard Deviations	
Between	0.0349679
Within	0.0262640
Between/Within	0.0437327

$\bar{x}$ chart: UCL = 11.8685, CL = 11.7579, LCL = 11.6472
MR chart: UCL = 0.1360, CL = 0.0416, LCL = 0
R chart: UCL = 0.1292, CL = 0.0611, LCL = 0

(d) Do you believe that the charts in part (c) are more informative than those in part (a)? Discuss why.

Yes, the charts in (c) are more informative than those in (a). The numerous out-of-control points on the $\bar{x}$ chart in (a) result from using the wrong source of variability to estimate sample variance. Recall that this is a cast part, and that multiple diameter measurements of any single part are likely to be very similar (text p. 246). With the three charts in (c), the correct variance estimate is used for each chart. In addition, more clear direction is provided by the within and between charts for trouble-shooting out-of-control signals.

(e) Provide a practical interpretation of the "within" chart.

The "within" chart is the usual R chart ($n > 1$). It describes the measurement variability within a sample (variability in diameter of a single casting). Though the nature of this process leads us to believe that the diameter at any location on a single casting does not change much, we should continue to monitor "within" to look for wear, damage, etc., in the wax mold.

5-71.   Consider the situation described in Exercise 5-70. A critical dimension (measured in μm) is of interest to the process engineer. Suppose that five fixed positions are used on each wafer (position 1 is the center) and that two consecutive wafers are selected from each batch. The data that result from several batches are shown here.

Lot Number	Wafer Number	Position 1	2	3	4	5	Lot Number	Wafer Number	Position 1	2	3	4	5
1	1	2.15	2.13	2.08	2.12	2.10	11	1	2.15	2.13	2.14	2.09	2.08
	2	2.13	2.10	2.04	2.08	2.05		2	2.11	2.13	2.10	2.14	2.10
2	1	2.02	2.01	2.06	2.05	2.08	12	1	2.03	2.06	2.05	2.01	2.00
	2	2.03	2.09	2.07	2.06	2.04		2	2.94	2.08	2.03	2.10	2.07
3	1	2.13	2.12	2.10	2.11	2.08	13	1	2.05	2.03	2.05	2.09	2.08
	2	2.03	2.08	2.03	2.09	2.07		2	2.08	2.01	2.03	2.04	2.10
4	1	2.04	2.01	2.10	2.11	2.09	14	1	2.08	2.04	2.05	2.01	2.08
	2	2.07	2.14	2.12	2.08	2.09		2	2.09	2.11	2.96	2.04	2.05
5	1	2.16	2.17	2.13	2.18	2.10	15	1	2.14	2.13	2.10	2.10	2.08
	2	2.17	2.13	2.10	2.09	2.13		2	2.13	2.10	2.09	2.13	2.15
6	1	2.04	2.06	1.97	2.10	2.08	16	1	2.06	2.08	2.05	2.03	2.09
	2	2.03	2.10	2.05	2.07	2.04		2	2.03	2.01	1.99	2.06	2.05
7	1	2.04	2.02	2.01	2.00	2.05	17	1	2.05	2.03	2.08	2.01	2.04
	2	2.06	2.04	2.03	2.08	2.10		2	2.06	2.05	2.03	2.05	2.00
8	1	2.13	2.10	2.10	2.15	2.13	18	1	2.03	2.08	2.04	2.00	2.03
	2	2.10	2.09	2.13	2.14	2.11		2	2.04	2.03	2.05	2.01	2.04
9	1	1.95	2.03	2.08	2.07	2.08	19	1	2.16	2.13	2.10	2.13	2.12
	2	2.01	2.03	2.06	2.05	2.04		2	2.13	2.15	2.18	2.19	2.13
10	1	2.04	2.08	2.09	2.10	2.91	20	1	2.06	2.03	2.04	2.09	2.10
	2	2.06	2.04	2.07	2.04	2.91		2	2.91	1.98	2.05	2.08	2.06

(a) What can you say about overall process capability?

MTB>Stat>Basic Statistics>Normality Test

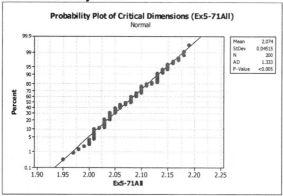

Although the *p*-value is very small, the plot points do fall along a straight line, with many repeated values. The wafer critical dimension is approximately normally distributed.

The natural tolerance limits (± 3 sigma above and below mean) are:
$\bar{x} = 2.074, s = 0.04515$

$\text{UNTL} = \bar{x} + 3s = 2.074 + 3(0.04515) = 2.209$

$\text{LNTL} = \bar{x} - 3s = 2.074 - 3(0.04515) = 1.939$

5-71 continued

(b) Can you construct control charts that allow within-wafer variability to be evaluated? To evaluate within-wafer variability, construct an R chart for each sample of 5 wafer positions (two wafers per lot number), for a total of 40 subgroups.

MTB>Stat>Control Charts>Variables Charts for Subgroups>R

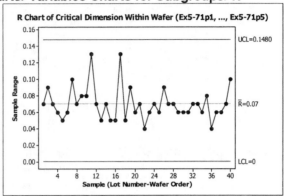

R chart: UCL = 0.1480, CL = 0.07, LCL = 0

The Range chart is in control, indicating that within-wafer variability is also in control.

(c) What control charts would you establish to evaluate variability between wafers? Set up these charts and use them to draw conclusions about the process.

To evaluate variability between wafers, set up Individuals and Moving Range charts where the x statistic is the average wafer measurement and the moving range is calculated between two wafer averages.

MTB>Stat>Control Charts>Variables Charts for Subgroups>I-MR-R/S (Between/Within)
Select "**I-MR-R/S Options, Estimate**" and choose R-bar method to estimate standard deviation

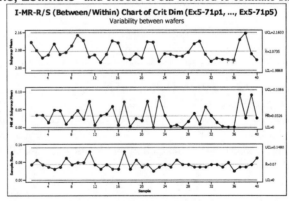

I-MR-R/S Standard Deviations of Ex5-71p1, ..., Ex5-71p5

```
Standard Deviations
Between          0.0255911
Within           0.0300946
Between/Within   0.0395043
```

5-71 continued

$\bar{x}$ chart: UCL = 2.1603, CL = 2.0735, LCL = 1.9868
MR chart: UCL =0.1066, CL = 0.0326, LCL = 0
R chart: UCL =0.148, CL = 0.07, LCL = 0

Both "between" control charts (Individuals and Moving Range) are in control, indicating that between-wafer variability is also in-control. The "within" chart (Range) is not required to evaluate variability between wafers.

(d) What control charts would you use to evaluate lot-to-lot variability? Set up these charts and use them to draw conclusions about lot-to-lot variability.

To evaluate lot-to-lot variability, three charts are needed: (1) lot average, (2) moving range between lot averages, and (3) range within a lot—the Minitab "between/within" control charts.

MTB>Stat>Control Charts>Variables Charts for Subgroups>I-MR-R/S (Between/Within)

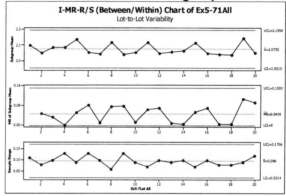

I-MR-R/S Standard Deviations of Ex5-71All
```
Standard Deviations
Between         0.0394733
Within          0.0311891
Between/Within  0.0503081
```

$\bar{x}$ chart: UCL = 2.1956, CL = 2.0735, LCL = 1.9515
MR chart: UCL =0.1500, CL = 0.0459, LCL = 0
R chart: UCL =0.1706, CL = 0.096, LCL = 0

All three control charts are in control, indicating that the lot-to-lot variability is also in-control.

CHAPTER 6

Control Charts for Attributes

Learning Objectives

After completing this chapter you should be able to:
1. Understand the statistical basis of attributes control charts
2. Know how to design attributes control charts
3. Know how to set up and use the p chart for fraction nonconforming
4. Know how to set up and use the np control chart for the number of nonconforming items
5. Know how to set up and use the c control chart for defects
6. Know how to set up and use the u control chart for defects per unit
7. Use attributes control charts with variable sample size
8. Understand the advantages and disadvantages of attributes versus variables control charts
9. Understand the rational subgroup concept for attributes control charts
10. Determine the average run length for attributes control charts

Important Terms and Concepts

Attribute data
Cause-and-effect diagram
Control chart for defects or nonconformities per unit or u chart
Control chart for nonconformities or c chart
Defect
Demerit systems for attribute data
Fraction defective
Nonconformities
Operating characteristic curve for the p chart
Standardized control charts
Variable sample size for attributes control chart

Average run length for attributes control charts
Choice between attributes and variables data
Control chart for fraction nonconforming or p chart

Control chart for number nonconforming or np chart
Defective
Design of attributes control charts
Fraction nonconforming
Operating characteristic curve for the c and u charts
Pareto chart
Time between occurrence control charts

Exercises

Notes:
1. For these solutions, we follow the MINITAB convention for determining whether a point is out of control. If a plot point is *within* the control limits, it is considered to be in control. If a plot point is *on* or *beyond* the control limits, it is considered to be out of control.
2. MINITAB defines some sensitizing rules for control charts differently than the standard rules. In particular, a run of n consecutive points on one side of the center line is defined as 9 points, not 8. This can be changed under **Tools>Options>Control Charts and Quality Tools>Define Tests**. Also fewer special cause tests are available for attributes control charts.

6-1. The data that follow give the number of nonconforming bearing and seal assemblies in samples of size 100. Construct a fraction nonconforming control chart for these data. If any points plot out of control, assume that assignable causes can be found and determine the revised control limits.

Sample Number	Number of Nonconforming Assemblies	Sample Number	Number of Nonconforming Assemblies
1	7	11	6
2	4	12	15
3	1	13	0
4	3	14	9
5	6	15	5
6	8	16	1
7	10	17	4
8	5	18	5
9	2	19	7
10	7	20	12

$$n = 100; \quad m = 20; \quad \sum_{i=1}^{m} D_i = 117; \quad \bar{p} = \frac{\sum_{i=1}^{m} D_i}{mn} = \frac{117}{20(100)} = 0.0585$$

$$\text{UCL}_p = \bar{p} + 3\sqrt{\frac{\bar{p}(1-\bar{p})}{n}} = 0.0585 + 3\sqrt{\frac{0.0585(1-0.0585)}{100}} = 0.1289 \qquad \text{(Equations 6-7, -8)}$$

$$\text{LCL}_p = \bar{p} - 3\sqrt{\frac{\bar{p}(1-\bar{p})}{n}} = 0.0585 - 3\sqrt{\frac{0.0585(1-0.0585)}{100}} = 0.0585 - 0.0704 \Rightarrow 0$$

6-1 continued

MTB>Stat>Control Charts>Attributes Charts>P

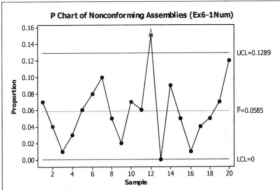

Test Results for P Chart of Ex6-1Num
TEST 1. One point more than 3.00 standard deviations from center line.
Test Failed at points: 12

p chart: UCL = 0.1289, CL = 0.0585, LCL = 0

Sample 12 is out-of-control, so remove from control limit calculation:

$$n = 100; \quad m = 19; \quad \sum_{i=1}^{m} D_i = 102; \quad \bar{p} = \sum_{i=1}^{m} D_i \bigg/ mn = 102/19(100) = 0.0537$$

$$\text{UCL}_p = 0.0537 + 3\sqrt{\frac{0.0537(1-0.0537)}{100}} = 0.1213 \qquad \text{(Equations 6-7, -8)}$$

$$\text{LCL}_p = 0.0537 - 3\sqrt{\frac{0.0537(1-0.0537)}{100}} = 0.0537 - 0.0676 \Rightarrow 0$$

MTB>Stat>Control Charts>Attributes Charts>P

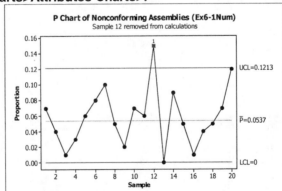

Test Results for P Chart of Ex6-1Num
TEST 1. One point more than 3.00 standard deviations from center line.
Test Failed at points: 12

p chart: UCL = 0.1213, CL = 0.0537, LCL = 0

6-3.   The following data represent the results of inspecting all units of a personal computer produced for the last 10 days. Does the process appear to be in control?

Day	Units Inspected	Nonconforming Units	Fraction Nonconforming
1	80	4	0.050
2	110	7	0.064
3	90	5	0.056
4	75	8	0.107
5	130	6	0.038
6	120	6	0.050
7	70	4	0.057
8	125	5	0.040
9	105	8	0.076
10	95	7	0.074

Note: There is an error in the table in the textbook. The Fraction Nonconforming for Day 5 should be 0.046.

$$m = 10; \quad \sum_{i=1}^{m} n_i = 1000; \quad \sum_{i=1}^{m} D_i = 60; \quad \bar{p} = \sum_{i=1}^{m} D_i \bigg/ \sum_{i=1}^{m} n_i = 60/1000 = 0.06$$

$$\text{UCL}_i = \bar{p} + 3\sqrt{\bar{p}(1-\bar{p})/n_i} \quad \text{and} \quad \text{LCL}_i = \max\{0, \bar{p} - 3\sqrt{\bar{p}(1-\bar{p})/n_i}\}$$

As an example, for $n = 80$:

$$\text{UCL}_1 = \bar{p} + 3\sqrt{\bar{p}(1-\bar{p})/n_1} = 0.06 + 3\sqrt{0.06(1-0.06)/80} = 0.1397$$

$$\text{LCL}_1 = \bar{p} - 3\sqrt{\bar{p}(1-\bar{p})/n_1} = 0.06 - 3\sqrt{0.06(1-0.06)/80} = 0.06 - 0.0797 \Rightarrow 0$$

(Equations 6-7, -8)

MTB>Stat>Control Charts>Attributes Charts>P

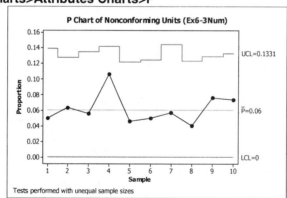

p chart: UCL = 0.1331, CL = 0.06, LCL = 0

The process appears to be in statistical control.

6-5. A process produces rubber belts in lots of size 2500. Inspection records on the last 20 lots reveal the following data.

Lot Number	Number of Nonconforming Belts	Lot Number	Number of Nonconforming Belts
1	230	11	456
2	435	12	394
3	221	13	285
4	346	14	331
5	230	15	198
6	327	16	414
7	285	17	131
8	311	18	269
9	342	19	221
10	308	20	407

(a) Compute trial control limits for a fraction nonconforming control chart.

$$\text{UCL} = \overline{p} + 3\sqrt{\overline{p}(1-\overline{p})/n} = 0.1228 + 3\sqrt{0.1228(1-0.1228)/2500} = 0.1425$$
$$\text{LCL} = \overline{p} - 3\sqrt{\overline{p}(1-\overline{p})/n} = 0.1228 - 3\sqrt{0.1228(1-0.1228)/2500} = 0.1031$$

(Equation 6-8)

MTB>Stat>Control Charts>Attributes Charts>P

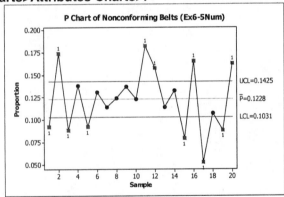

Test Results for P Chart of Ex6-5Num
TEST 1. One point more than 3.00 standard deviations from center line.
Test Failed at points: 1, 2, 3, 5, 11, 12, 15, 16, 17, 19, 20

p chart: UCL = 0.1425, CL = 0.1228, LCL = 0.1031

(b) If you wanted to set up a control chart for controlling future production, how would you use these data to obtain the center line and control limits for the chart?

So many subgroups are out of control (11 of 20) that the data should not be used to establish control limits for future production. Instead, the process should be investigated for causes of the wild swings in p.

6-7. A control chart indicates that the current process fraction nonconforming is 0.02. If 50 items are inspected each day, what is the probability of detecting a shift in the fraction nonconforming to 0.04 on the first day after the shift? By the end of the third day following the shift?

$$\bar{p} = 0.02; n = 50$$

$$\text{UCL} = \bar{p} + 3\sqrt{\bar{p}(1-\bar{p})/n} = 0.02 + 3\sqrt{0.02(1-0.02)/50} = 0.0794 \qquad \text{(Equation 6-8)}$$

$$\text{LCL} = \bar{p} - 3\sqrt{\bar{p}(1-\bar{p})/n} = 0.02 - 3\sqrt{0.02(1-0.02)/50} = 0.02 - 0.0594 \Rightarrow 0$$

Since $p_{new} = 0.04 < 0.1$ and $n = 50$ is "large", use the Poisson approximation to the binomial with $\lambda = np_{new} = 50(0.04) = 2.00$. (text page 78)

$$\begin{aligned}
\text{Pr}\{\text{detect}|\text{shift}\} \\
&= 1 - \text{Pr}\{\text{not detect}|\text{shift}\} \\
&= 1 - \beta \\
&= 1 - [\text{Pr}\{D < n\text{UCL} \mid \lambda\} - \text{Pr}\{D \le n\text{LCL} \mid \lambda\}] \qquad \text{(Equation 6-15)} \\
&= 1 - \text{Pr}\{D < 50(0.0794) \mid 2\} + \text{Pr}\{D \le 50(0) \mid 2\} \\
&= 1 - \text{POI}(3,2) + \text{POI}(0,2) = 1 - 0.857 + 0.135 = 0.278
\end{aligned}$$

where POI($\cdot$) is the cumulative Poisson distribution.

$$\text{Pr}\{\text{detected by 3rd sample}\} = 1 - \text{Pr}\{\text{detected after 3rd}\} = 1 - (1 - 0.278)^3 = 0.624$$

6-9.   Diodes used on printed circuit boards are produced in lots of size 1000. We wish to control the process producing these diodes by taking samples of size 64 from each lot. If the nominal value of the fraction nonconforming is $p = 0.10$, determine the parameters of the appropriate control chart. To what level must the fraction nonconforming increase to make the β-risk equal to 0.50? What is the minimum sample size that would give a positive lower control limit for this chart?

$$\bar{p} = 0.10; n = 64$$

$$\text{UCL} = \bar{p} + 3\sqrt{\bar{p}(1-\bar{p})/n} = 0.10 + 3\sqrt{0.10(1-0.10)/64} = 0.2125 \qquad \text{(Equation 6-8)}$$

$$\text{LCL} = \bar{p} - 3\sqrt{\bar{p}(1-\bar{p})/n} = 0.10 - 3\sqrt{0.10(1-0.10)/64} = 0.10 - 0.1125 \Rightarrow 0$$

$$\beta = \Pr\{D < n\text{UCL} \mid p\} - \Pr\{D \leq n\text{LCL} \mid p\}$$
$$= \Pr\{D < 64(0.2125) \mid p\} - \Pr\{D \leq 64(0) \mid p\} \text{ (Equation 6-15)}$$
$$= \Pr\{D < 13.6) \mid p\} - \Pr\{D \leq 0 \mid p\}$$

p	Pr{D ≤ 13|p}	Pr{D ≤ 0|p}	β
0.05	0.999999	0.037524	0.962475
0.10	0.996172	0.001179	0.994993
0.20	0.598077	0.000000	0.598077
0.21	**0.519279**	**0.000000**	**0.519279**
0.22	0.44154	0.000000	0.44154
0.215	0.480098	0.000000	0.480098
0.212	0.503553	0.000000	0.503553

Assuming $L = 3$ sigma control limits,

$$n > \frac{(1-p)}{p} L^2$$

$$> \frac{(1-0.10)}{0.10}(3)^2 \quad \text{(Equation 6-12)}$$

$$> 81$$

6-11.   A control chart for the fraction nonconforming is to be established using a center line of $p = 0.10$. What sample size is required if we wish to detect a shift in the process fraction nonconforming to 0.20 with probability 0.50?

$p = 0.10$; $p_{new} = 0.20$; desire Pr{detect} $= 0.50$; assume $L = 3$ sigma control limits

$\delta = p_{new} - p = 0.20 - 0.10 = 0.10$ (Equation 6-10)

$$n = \left(\frac{L}{\delta}\right)^2 p(1-p) = \left(\frac{3}{0.10}\right)^2 (0.10)(1-0.10) = 81$$

6-13.   A process is being controlled with a fraction nonconforming control chart. The process average has been shown to be 0.07. Three-sigma control limits are used, and the procedure calls for taking daily samples of 400 items.

(a) Calculate the upper and lower control limits.

$\bar{p} = 0.07$; $k = 3$ sigma control limits; $n = 400$

$\text{UCL} = \bar{p} + 3\sqrt{p(1-p)/n} = 0.07 + 3\sqrt{0.07(1-0.07)/400} = 0.108$ (Equation 6-8)

$\text{LCL} = \bar{p} - 3\sqrt{p(1-p)/n} = 0.07 - 3\sqrt{0.07(1-0.07)/400} = 0.032$

(b) If the process average should suddenly shift to 0.10, what is the probability that the shift would be detected on the first subsequent sample?

$np_{new} = 400(0.10) => 40$, so use the normal approximation to the binomial (text page 78)

Pr{detect on 1st sample} $= 1 - $Pr{not detect on 1st sample}

$$= 1 - \beta$$

$$= 1 - [\text{Pr}\{\hat{p} < \text{UCL} \mid p\} - \text{Pr}\{\hat{p} \leq \text{LCL} \mid p\}]$$

$$= 1 - \Phi\left(\frac{\text{UCL} - p}{\sqrt{p(1-p)/n}}\right) + \Phi\left(\frac{\text{LCL} - p}{\sqrt{p(1-p)/n}}\right)$$

(Equation 6-15)

$$= 1 - \Phi\left(\frac{0.108 - 0.1}{\sqrt{0.1(1-0.1)/400}}\right) + \Phi\left(\frac{0.032 - 0.1}{\sqrt{0.1(1-0.1)/400}}\right)$$

$$= 1 - \Phi(0.533) + \Phi(-4.533)$$

$$= 1 - 0.703 + 0.000$$

$$= 0.297$$

(c) What is the probability that the shift in part (b) would be detected on the first or second sample taken after the shift?

Pr{detect on 1st or 2nd sample}

$= $ Pr{detect on 1st} $ + $ Pr{not on 1st}$\times$Pr{detect on 2nd}

$= 0.297 + (1 - 0.297)(0.297)$

$= 0.506$

6-15. A control chart is used to control the fraction nonconforming for a plastic part manufactured in an injection molding process. Ten subgroups yield the following data:

Sample Number	Sample Size	Number Nonconforming
1	100	10
2	100	15
3	100	31
4	100	18
5	100	24
6	100	12
7	100	23
8	100	15
9	100	8
10	100	8

(a) Set up a control chart for the number nonconforming in samples of $n = 100$.

$$m = 10; \quad n = 100; \quad \sum_{i=1}^{10} D_i = 164; \quad \overline{p} = \sum_{i=1}^{10} D_i \Big/ (mn) = 164/[10(100)] = 0.164; \quad n\overline{p} = 16.4$$

$$\text{UCL} = n\overline{p} + 3\sqrt{n\overline{p}(1-\overline{p})} = 16.4 + 3\sqrt{16.4(1-0.164)} = 27.51$$

$$\text{LCL} = n\overline{p} - 3\sqrt{n\overline{p}(1-\overline{p})} = 16.4 - 3\sqrt{16.4(1-0.164)} = 5.292$$

MTB>Stat>Control Charts>Attributes Charts>NP

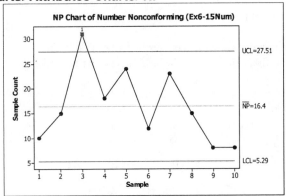

Test Results for NP Chart of Ex6-15Num
TEST 1. One point more than 3.00 standard deviations from center line.
Test Failed at points: 3

np chart: UCL = 27.51, CL = 16.4, LCL = 5.29

6-15 continued

Recalculate control limits less sample 3:

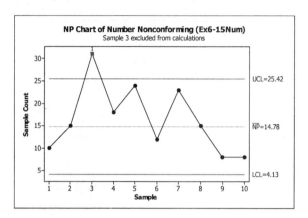

Test Results for NP Chart of Ex6-15Num
TEST 1. One point more than 3.00 standard deviations from center line.
Test Failed at points: 3

---

np chart: UCL = 25.42, CL = 14.78, LCL = 4.13

These limits will be used for future production.

(b) For the chart established in part (a), what is the probability of detecting a shift in the process fraction nonconforming to 0.30 on the first sample after the shift has occurred?

$p_{new} = 0.30$. Since $p = 0.30$ is not too far from 0.50, and $n = 100 > 10$, the normal approximation to the binomial can be used.

$\Pr\{\text{detect on 1st}\} = 1 - \Pr\{\text{not detect on 1st}\}$

$$= 1 - \beta$$

$$= 1 - [\Pr\{D < \text{UCL} \mid p\} - \Pr\{D \leq \text{LCL} \mid p\}]$$

$$= 1 - \Phi\left(\frac{\text{UCL} + 1/2 - np}{\sqrt{np(1-p)}}\right) + \Phi\left(\frac{\text{LCL} - 1/2 - np}{\sqrt{np(1-p)}}\right)$$

$$= 1 - \Phi\left(\frac{25.42 + 0.5 - 30}{\sqrt{30(1-0.3)}}\right) + \Phi\left(\frac{4.13 - 0.5 - 30}{\sqrt{30(1-0.3)}}\right)$$

$$= 1 - \Phi(-0.8903) + \Phi(-5.7544)$$

$$= 1 - (0.187) + (0.000)$$

$$= 0.813$$

6-17.   (a) A control chart for the number nonconforming is to be established, based on samples of size 400. To start the control chart, 30 samples were selected and the number nonconforming in each sample determined, yielding $\sum_{i=1}^{30} D_i = 1200$. What are the parameters of the *np* chart?

$$\bar{p} = \sum_{i=1}^{m} D_i \bigg/ (mn) = 1200 \big/ [30(400)] = 0.10; \quad n\bar{p} = 400(0.10) = 40$$

$$\mathrm{UCL}_{np} = n\bar{p} + 3\sqrt{n\bar{p}(1-\bar{p})} = 40 + 3\sqrt{40(1-0.10)} = 58 \qquad \text{(Equation 13)}$$

$$\mathrm{LCL}_{np} = n\bar{p} - 3\sqrt{n\bar{p}(1-\bar{p})} = 40 - 3\sqrt{40(1-0.10)} = 22$$

(b) Suppose the process average fraction nonconforming shifted to 0.15. What is the probability that the shift would be detected on the first subsequent sample?

$np_{new} = 400\,(0.15) = 60 > 15$, so use the normal approximation to the binomial (text page 78)

$$\Pr\{\text{detect on 1st sample} \mid p\} = 1 - \Pr\{\text{not detect on 1st sample} \mid p\}$$

$$= 1 - \beta$$

$$= 1 - [\Pr\{D < \mathrm{UCL} \mid np\} - \Pr\{D \le \mathrm{LCL} \mid np\}]$$

$$= 1 - \Phi\left(\frac{\mathrm{UCL} + 1/2 - np}{\sqrt{np(1-p)}} \right) + \Phi\left(\frac{\mathrm{LCL} - 1/2 - np}{\sqrt{np(1-p)}} \right) \quad \text{(Equation 6-15)}$$

$$= 1 - \Phi\left(\frac{58 + 0.5 - 60}{\sqrt{60(1-0.15)}} \right) + \Phi\left(\frac{22 - 0.5 - 60}{\sqrt{60(1-0.15)}} \right)$$

$$= 1 - \Phi(-0.210) + \Phi(-5.39)$$

$$= 1 - 0.417 + 0.000$$

$$= 0.583$$

6-19.   Consider the control chart designed in Exercise 6-18. Find the average run length to detect a shift to a fraction nonconforming of 0.15. (**Note:** There is an error in the textbook. This is a continuation of Exercise 6-17, not 6-18.)

From 6-17(b), $1 - \beta = 0.583$

$\mathrm{ARL}_1 = 1/(1 - \beta) = 1/(0.583) = 1.715 \cong 2$

6-21.   A maintenance group improves the effectiveness of its repair work by monitoring the number of maintenance requests that require a second call to complete the repair. Twenty weeks of data are available.

Week	Total Requests	Second Visit Required	Week	Total Requests	Second Visit Required
1	200	6	11	100	1
2	250	8	12	100	0
3	250	9	13	100	1
4	250	7	14	200	4
5	200	3	15	200	5
6	200	4	16	200	3
7	150	2	17	200	10
8	150	1	18	200	4
9	150	0	19	250	7
10	150	2	20	250	6

(a) Find trial control limits for this process.

   For a p chart with variable sample size: $\bar{p} = \sum_i D_i / \sum_i n_i = 83/3750 = 0.0221$ and control limits are at $\bar{p} \pm 3\sqrt{\bar{p}(1-\bar{p})/n_i}$ (text page 280).

n_i	[LCL$_i$, UCL$_i$]
100	[0, 0.0662]
150	[0, 0.0581]
200	[0, 0.0533]
250	[0, 0.0500]

MTB>Stat>Control Charts>Attributes Charts>P

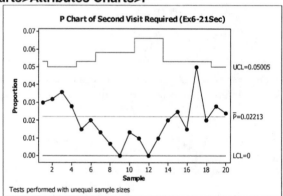

Process is in statistical control.

(b) Design a control chart for controlling future production.

   There are two approaches for controlling future production. The first approach would be to plot $\hat{p}_i$ and use constant limits unless there is a different size sample or a plot point near a control limit. In those cases, calculate the exact control limits by $\bar{p} \pm 3\sqrt{\bar{p}(1-\bar{p})/n_i} = 0.0221 \pm 3\sqrt{0.0216/n_i}$. The second approach, preferred in many cases, would be to construct standardized control limits with control limits at ± 3, and to plot $Z_i = (\hat{p}_i - 0.0221)/\sqrt{0.0221(1-0.0221)/n_i}$.

6-23. Construct a standardized control chart for the data in Exercise 6-21.

$$z_i = (\hat{p}_i - \overline{p})\big/\sqrt{\overline{p}(1-\overline{p})/n_i} = (\hat{p}_i - 0.0221)\big/\sqrt{0.0216/n_i} \quad \text{(Equation 6-14)}$$

MTB>Stat>Control Charts>Variables Charts for Individuals>Individuals (Ex6-23zi)

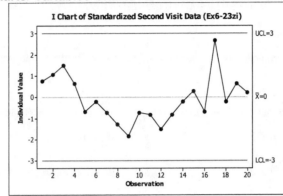

"Standardized" chart: UCL = 3, CL = 0, LCL = -3

Process is in statistical control.

6-25. A fraction nonconforming control chart has center line 0.01, UCL = 0.0399, LCL = 0, and $n = 100$. If three-sigma limits are used, find the smallest sample size that would yield a positive lower control limit.

UCL = 0.0399; $\overline{p}$ = CL = 0.01; LCL = 0; $n = 100$

$$n > \left(\frac{1-p}{p}\right)L^2$$

$$> \left(\frac{1-0.01}{0.01}\right)3^2 \quad \text{(Equation 6-12)}$$

$$> 891$$

$$\geq 892$$

6-27.   A fraction nonconforming control chart with n = 400 has the following parameters: UCL = 0.0809, Center line = 0.0500, LCL = 0.0191.

(a) Find the width of the control limits in standard deviation units.

$$n = 400; \quad \text{UCL} = 0.0809; \quad p = \text{CL} = 0.0500; \quad \text{LCL} = 0.0191$$

$$0.0809 = 0.05 + L\sqrt{0.05(1-0.05)/400} = 0.05 + L(0.0109)$$

$$L = 2.8349$$

(b) What would be the corresponding parameters for an equivalent control chart based on the number nonconforming?

$$\text{CL} = np = 400(0.05) = 20$$

$$\text{UCL} = np + 2.8349\sqrt{np(1-p)} = 20 + 2.8349\sqrt{20(1-0.05)} = 32.36 \quad \text{(Equation 6-13)}$$

$$\text{LCL} = np - 2.8349\sqrt{np(1-p)} = 20 - 2.8349\sqrt{20(1-0.05)} = 7.64$$

(c) What is the probability that a shift in the process fraction nonconforming to 0.0300 will be detected on the first sample following the shift?

$n = 400$ is large and $p = 0.05 < 0.1$, use Poisson approximation to binomial (text page 78)

Pr{detect shift to 0.03 on 1st sample}

$$= 1 - \text{Pr\{not detect\}}$$

$$= 1 - \beta$$

$$= 1 - [\text{Pr}\{D < \text{UCL} \mid \lambda\} - \text{Pr}\{D \le \text{LCL} \mid \lambda\}] \quad \text{(Equation 6-15)}$$

$$= 1 - \text{Pr}\{D < 32.36 \mid 12\} + \text{Pr}\{D \le 7.64 \mid 12\}$$

$$= 1 - \text{POI}(32,12) + \text{POI}(7,12)$$

$$= 1 - 1.0000 + 0.0895$$

$$= 0.0895$$

where POI($\cdot$) is the cumulative Poisson distribution.

6-29. A fraction nonconforming control chart is to be established with a center line of 0.01 and two-sigma control limits.

(a) How large should the sample size be if the lower control limit is to be nonzero?

$p = 0.01; \ L = 2$

$$n > \left(\frac{1-p}{p} \right) L^2$$

$$> \left(\frac{1-0.01}{0.01} \right) 2^2 \ \text{(Equation 6-12)}$$

$$> 396$$

$$\geq 397$$

(b) How large should the sample size be if we wish the probability of detecting a shift to 0.04 to be 0.50?

$\delta = 0.04 - 0.01 = 0.03$

$$n = \left(\frac{L}{\delta} \right)^2 p(1-p) = \left(\frac{2}{0.03} \right)^2 (0.01)(1-0.01) = 44 \ \text{(Equation 6-10)}$$

6-31. A process that produces bearing housings is controlled with a fraction nonconforming control chart, using sample size $n = 100$ and a center line $\bar{p} = 0.02$.

(a) Find the three-sigma limits for this chart.

$n = 100; \quad \bar{p} = 0.02$

$$\text{UCL} = \bar{p} + 3\sqrt{\bar{p}(1-\bar{p})/n} = 0.02 + 3\sqrt{0.02(1-0.02)/100} = 0.062$$

$$\text{LCL} = \bar{p} - 3\sqrt{\bar{p}(1-\bar{p})/n} = 0.02 - 3\sqrt{0.02(1-0.02)/100} \Rightarrow 0$$

(Equation 6-8)

(b) Analyze the ten new samples ($n = 100$) shown here for statistical control. What conclusions can you draw about the process now?

Sample Number	Number Nonconforming	Sample Number	Number Nonconforming
1	5	6	1
2	2	7	2
3	3	8	6
4	8	9	3
5	4	10	4

MTB>Stat>Control Charts>Attributes Charts>P

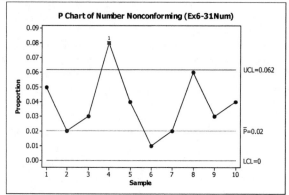

Test Results for P Chart of Ex6-31Num
```
TEST 1. One point more than 3.00 standard deviations from center line.
Test Failed at points:   4
```

p chart: UCL = 0.062, CL = 0.02, LCL = 0

Sample 4 exceeds the upper control limit, signaling a potentially unstable process.

$\bar{p} = 0.038$ and $\hat{\sigma}_p = 0.0191$

6-33. Consider the fraction nonconforming control chart in Exercise 6-4. Find the equivalent *np* chart.

Day	Nonconforming Units	Day	Nonconforming Units
1	3	11	2
2	2	12	4
3	4	13	1
4	2	14	3
5	5	15	6
6	2	16	0
7	1	17	1
8	2	18	2
9	0	19	3
10	5	20	2

$n = 150; \quad m = 20; \quad \sum D = 50; \quad \overline{p} = 0.0167$

$\text{CL} = n\overline{p} = 150(0.0167) = 2.505$

$\text{UCL} = n\overline{p} + 3\sqrt{n\overline{p}(1-\overline{p})} = 2.505 + 3\sqrt{2.505(1-0.0167)} = 7.213$ (Equation 6-13)

$\text{LCL} = n\overline{p} - 3\sqrt{n\overline{p}(1-\overline{p})} = 2.505 - 4.708 \Rightarrow 0$

MTB>Stat>Control Charts>Attributes Charts>NP

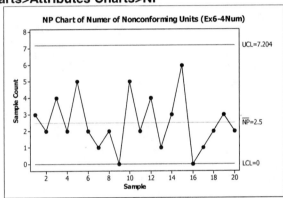

np chart: UCL = 7.204, CL = 2.5, LCL = 0

The process is in control; results are the same as for the *p* chart.

6-35. Construct a standardized control chart for the data in Exercise 6-3.

Day	Units Inspected	Nonconforming Units	Fraction Nonconforming
1	80	4	0.050
2	110	7	0.064
3	90	5	0.056
4	75	8	0.107
5	130	6	0.038
6	120	6	0.050
7	70	4	0.057
8	125	5	0.040
9	105	8	0.076
10	95	7	0.074

$$\overline{p} = 0.06$$

$$z_i = (\hat{p}_i - 0.06)\big/\sqrt{0.06(1-0.06)/n_i} = (\hat{p}_i - 0.06)\big/\sqrt{0.0564/n_i}$$

MTB > Stat > Control Charts > Variables Charts for Individuals > Individuals

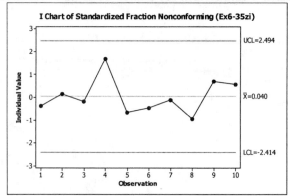

"Standardized" chart: UCL = 2.494, CL = 0, LCL = -2.414

The process is in control; results are the same as for the p chart.

6-37. A paper mill uses a control chart to monitor the imperfection in finished rolls of paper. Production output is inspected for 20 days, and the resulting data are shown here. Use these data to set up a control chart for nonconformities per roll of paper. Does the process appear to be in statistical control? What center line and control limits would you recommend for controlling current production?

Day	Number of Rolls Produced	Total Number of Imperfections	Day	Number of Rolls Produced	Total Number of Imperfections
1	18	12	11	18	18
2	18	14	12	18	14
3	24	20	13	18	9
4	22	18	14	20	10
5	22	15	15	20	14
6	22	12	16	20	13
7	20	11	17	24	16
8	20	15	18	24	18
9	20	12	19	22	20
10	20	10	20	21	17

$$CL = \bar{u} = 0.7007$$

$$UCL_i = \bar{u} + 3\sqrt{\bar{u}/n_i} = 0.7007 + 3\sqrt{0.7007/n_i} \quad \text{(Equation 6-19 and text page 299)}$$

$$LCL_i = \bar{u} - 3\sqrt{\bar{u}/n_i} = 0.7007 - 3\sqrt{0.7007/n_i}$$

n_i	[LCL$_i$, UCL$_i$]
18	[0.1088, 1.2926]
20	[0.1392, 1.2622]
21	[0.1527, 1.2487]
22	[0.1653, 1.2361]
24	[0.1881, 1.2133]

MTB > Stat > Control Charts > Attributes Charts > U

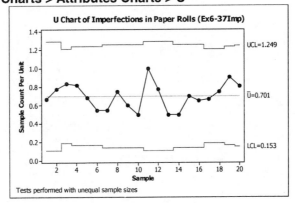

Process is in control, with no out-of-control signals.

6-39.   Continuation of Exercise 6-37. Consider the papermaking process in Exercise 6-37. Set up a standardized u chart for this process.

$$z_i = (u_i - \overline{u}) \big/ \sqrt{\overline{u}/n_i} = (u_i - 0.7007) \big/ \sqrt{0.7007/n_i} \quad \text{(Equation 6-20)}$$

MTB>Stat>Control Charts>Variables Charts for Individuals>Individuals

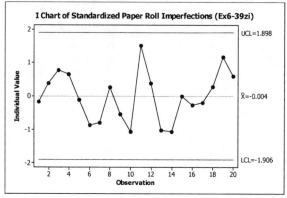

"Standardized" chart: UCL = 1.898, CL = 0.004, LCL = -1.906

Process is in control, with no out-of-control signals or unusual patterns.

6-41. The following data represent the number of nonconformities per 1000 meters in telephone cable. From analysis of these data, would you conclude that the process is in statistical control? What control procedure would you recommend for future production?

Sample Number	Number of Nonconformities	Sample Number	Number of Nonconformities
1	1	12	6
2	1	13	9
3	3	14	11
4	7	15	15
5	8	16	8
6	10	17	3
7	5	18	6
8	13	19	7
9	0	20	4
10	19	21	9
11	24	22	20

$$CL = \bar{c} = 8.59; \quad UCL = \bar{c} + 3\sqrt{\bar{c}} = 8.59 + 3\sqrt{8.59} = 17.38; \quad LCL = \bar{c} - 3\sqrt{\bar{c}} = 8.59 - 3\sqrt{8.59} \Rightarrow 0$$
(Equation 6-17)

MTB>Stat>Control Charts>Attributes Charts>C

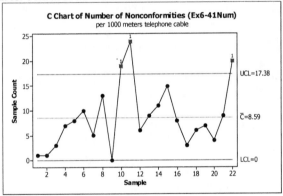

Test Results for C Chart of Ex6-41Num
```
TEST 1. One point more than 3.00 standard deviations from center line.
Test Failed at points:  10, 11, 22
```

Process is not in statistical control; three subgroups exceed the UCL. Exclude subgroups 10, 11 and 22, then re-calculate the control limits. Subgroup 15 will then be out of control and should also be excluded.

6-41 continued

$$CL = \bar{c} = 6.17; \quad UCL = \bar{c} + 3\sqrt{\bar{c}} = 6.17 + 3\sqrt{6.17} = 13.62; \quad LCL \Rightarrow 0$$

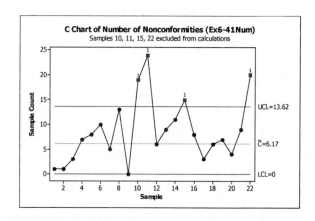

Test Results for C Chart of Ex6-41Num
TEST 1. One point more than 3.00 standard deviations from center line.
Test Failed at points: 10, 11, 15, 22

6-43.   Consider the data in Exercise 6-41. Suppose a new inspection unit is defined as 2500 m of wire.

(a) What are the center line and control limits for a control chart for monitoring future production based on the total number of nonconformities in the new inspection unit?

> The new inspection unit is $n = 2500/1000 = 2.5$ of the old unit. A c chart of the total number of nonconformities per inspection unit is appropriate.

$$CL = n\bar{c} = 2.5(6.17) = 15.43$$

$$UCL = n\bar{c} + 3\sqrt{n\bar{c}} = 15.43 + 3\sqrt{15.43} = 27.21 \text{ (Equation 6-17)}$$

$$LCL = n\bar{c} - 3\sqrt{n\bar{c}} = 15.43 - 3\sqrt{15.43} = 3.65$$

> The plot point, $\hat{c}$, is the total number of nonconformities found while inspecting a sample 2500m in length.

(b) What are the center line and control limits for a control chart for average nonconformities per unit used to monitor future production?

> The sample is $n = 1$ new inspection units. A u chart of average nonconformities per inspection unit is appropriate.

$$CL = \bar{u} = \frac{\text{total nonconformities}}{\text{total inspection units}} = \frac{111}{(18 \times 1000)/2500} = 15.42$$

$$UCL = \bar{u} + 3\sqrt{\bar{u}/n} = 15.42 + 3\sqrt{15.42/1} = 27.20 \qquad \text{(Equation 6-19)}$$

$$LCL = \bar{u} - 3\sqrt{\bar{u}/n} = 15.42 - 3\sqrt{15.42/1} = 3.64$$

> The plot point, $\hat{u}$, is the average number of nonconformities found in 2500m, and since $n = 1$, this is the same as the total number of nonconformities.

6-45.   Find the three-sigma control limits for:

(a) a c chart with process average equal to four nonconformities.

$$CL = \bar{c} = 4$$

$$UCL = \bar{c} + 3\sqrt{\bar{c}} = 4 + 3\sqrt{4} = 10 \text{ (Equation 6-17)}$$

$$LCL = \bar{c} - 3\sqrt{\bar{c}} = 4 - 3\sqrt{4} \Rightarrow 0$$

(b) a u chart with c = 4 and n = 4.

$$c = 4; \quad n = 4$$

$$CL = \bar{u} = c/n = 4/4 = 1$$

$$UCL = \bar{u} + 3\sqrt{\bar{u}/n} = 1 + 3\sqrt{1/4} = 2.5 \quad \text{(Equations 6-18 and 6-19)}$$

$$LCL = \bar{u} - 3\sqrt{\bar{u}/n} = 1 - 3\sqrt{1/4} \Rightarrow 0$$

6-47.   Find the three-sigma control limits for:

(a) a c chart with process average equal to nine nonconformities.

$$CL = \bar{c} = 9$$

$$UCL = \bar{c} + 3\sqrt{\bar{c}} = 9 + 3\sqrt{9} = 18 \text{ (Equation 6-17)}$$

$$LCL = \bar{c} - 3\sqrt{\bar{c}} = 9 - 3\sqrt{9} = 0$$

(b) a u chart with c = 16 and n = 4.

$$c = 16; \quad n = 4$$

$$CL = \bar{u} = c/n = 16/4 = 4$$

$$UCL = \bar{u} + 3\sqrt{\bar{u}/n} = 4 + 3\sqrt{4/4} = 7 \quad \text{(Equations 6-18 and 6-19)}$$

$$LCL = \bar{u} - 3\sqrt{\bar{u}/n} = 4 - 3\sqrt{4/4} = 1$$

6-49.   Find 0.975 and 0.025 probability limits for a control chart for nonconformities when c = 7.6.

Using the cumulative Poisson distribution:

x	$Pr\{D \le x \mid c = 7.6\}$
2	0.019
3	0.055
12	0.954
13	0.976

For the c chart, UCL = 13 and LCL = 2. As a comparison, the normal distribution gives

$$UCL = \bar{c} + z_{0.975}\sqrt{\bar{c}} = 7.6 + 1.96\sqrt{7.6} = 13.00$$

$$LCL = \bar{c} - z_{0.025}\sqrt{\bar{c}} = 7.6 - 1.96\sqrt{7.6} = 2.20$$

6-51.   The number of workmanship nonconformities observed in the final inspection of disk-drive assemblies has been tabulated as shown here. Does the process appear to be in control?

Day	Number of Assemblies Inspected	Total Number of Imperfections	Day	Number of Assemblies Inspected	Total Number of Imperfections
1	2	10	6	4	24
2	4	30	7	2	15
3	2	18	8	4	26
4	1	10	9	3	21
5	3	20	10	1	8

u chart with control limits based on each sample size: $\bar{u} = 7$; $UCL_i = 7 + 3\sqrt{7/n_i}$; $LCL_i = 7 - 3\sqrt{7/n_i}$

MTB>Stat>Control Charts>Attributes Charts>U

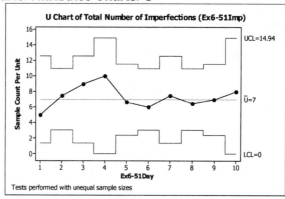

The process is in statistical control.

6-53.   A textile mill wishes to establish a control procedure on flaws in towels it manufactures. Using an inspection unit of 50 units, past inspection data show that 100 previous inspection units had 850 total flaws. What type of control chart is appropriate? Design the control chart such that it has two-sided probability control limits of $\alpha = 0.06$, approximately. Give the center line and control limits.

A c chart with one inspection unit equal to 50 manufacturing units is appropriate. $\bar{c} = 850/100 = 8.5$.
From the cumulative Poisson distribution:

x	Pr{D ≤ x	c = 8.5}
3	0.030
13	0.949
14	0.973

LCL = 3 and UCL = 13. For comparison, the normal distribution gives
$$UCL = \bar{c} + z_{0.97}\sqrt{\bar{c}} = 8.5 + 1.88\sqrt{8.5} = 13.98$$
$$LCL = \bar{c} + z_{0.03}\sqrt{\bar{c}} = 8.5 - 1.88\sqrt{8.5} = 3.02$$

6-55. Assembled portable television sets are subjected to a final inspection for surface defects. A total procedure is established based on the requirement that if the average number of nonconformities per unit is 4.0, the probability of concluding that the process is in control will be 0.99. There is to be no lower control limit. What is the appropriate type of control chart and what is the required upper control limit?

$\bar{u} = 4.0$ average number of nonconformities/unit. Desire $\alpha = 0.99$. Use the cumulative Poisson distribution to determine the UCL.

Using **EXCEL**

Ex6-55X	Ex6-55alpha
0	=POISSON(0,4,TRUE)
1	=POISSON(1,4,TRUE)
2	=POISSON(2,4,TRUE)
3	=POISSON(3,4,TRUE)
4	=POISSON(4,4,TRUE)
5	=POISSON(5,4,TRUE)
6	=POISSON(A8,4,TRUE)
7	=POISSON(A9,4,TRUE)
8	=POISSON(A10,4,TRUE)
9	=POISSON(A11,4,TRUE)
10	=POISSON(A12,4,TRUE)
11	=POISSON(A13,4,TRUE)

Ex6-55X	Ex6-55alpha
0	0.02
1	0.09
2	0.24
3	0.43
4	0.63
5	0.79
6	0.89
7	0.95
8	0.98
9	0.99
10	1.00
11	1.00

An UCL = 9 will give a probability of 0.99 of concluding the process is in control, when in fact it is.

6-57. Consider the situation described in Exercise 6-56.

Exercise 6-56: A control chart is to be established on a process producing refrigerators. The inspection unit is one refrigerator, and a common chart for nonconformities is to be used. As preliminary data, 16 nonconformities were counted in inspecting 30 refrigerators.

(a) Find two-sigma control limits and compare these with the control limits found in part (a) of Exercise 6-56.

Use a c chart for nonconformities with an inspection unit $n = 1$ refrigerator.

$$\sum_i D_i = 16 \text{ in } 30 \text{ refrigerators}; \quad \bar{c} = 16/30 = 0.533$$

3-sigma limits are $\bar{c} \pm 3\sqrt{\bar{c}} = 0.533 \pm 3\sqrt{0.533} = [0, 2.723]$

(b) Find the α-risk for the control chart with two-sigma control limits and compare with the results of part (b) of Exercise 6-56.

$$\alpha = \Pr\{D < \text{LCL} \mid c\} + \Pr\{D > \text{UCL} \mid c\}$$

$$= \Pr\{D < 0 \mid 0.533\} + \left[1 - \Pr\{D \le 2.72 \mid 0.533\}\right]$$

$$= 0 + \left[1 - \text{POI}(2, 0.533)\right] \qquad \text{(text page 158)}$$

$$= 1 - 0.983$$

$$= 0.017$$

where POI($\cdot$) is the cumulative Poisson distribution.

(c) Find the β-risk for $c = 2.0$ for the chart with two-sigma control limits and compare with the results of part (c) of Exercise 6-56.

$$\beta = \Pr\{\text{not detecting shift}\}$$

$$= \Pr\{D < \text{UCL} \mid c\} - \Pr\{D \le \text{LCL} \mid c\}$$

$$= \Pr\{D < 2.72 \mid 2.0\} - \Pr\{D \le 0 \mid 2.0\} \qquad \text{(Equation 6-26)}$$

$$= \text{POI}(2, 2) - \text{POI}(0, 2)$$

$$= 0.6767 - 0.1353$$

$$= 0.5414$$

where POI($\cdot$) is the cumulative Poisson distribution.

(d) Find the ARL if $c = 2.0$ and compare with the ARL found in part (d) of Exercise 6-56.

$$\text{ARL}_1 = \frac{1}{1 - \beta} = \frac{1}{1 - 0.541} = 2.18 \approx 2 \quad \text{(text page 287)}$$

6-59.   A control chart for nonconformities is maintained on a process producing desk calculators. The inspection unit is defined as two calculators. The average number of nonconformities per machine when the process is in control is estimated to be two.

(a) Find the appropriate three-sigma control limits for this size inspection unit.

$\bar{u}$ = average # nonconformities/calculator = 2

c chart with $\bar{c} = \bar{u} \times n = 2(2) = 4$ nonconformities/inspection unit

$CL = \bar{c} = 4$

$UCL = \bar{c} + k\sqrt{\bar{c}} = 4 + 3\sqrt{4} = 10$

$LCL = \bar{c} - k\sqrt{\bar{c}} = 4 - 3\sqrt{4} \Rightarrow 0$

(b) What is the probability of type I error for this control chart?

Type I error =

$$\begin{aligned} \alpha &= \Pr\{D < LCL\,|\,\bar{c}\} + \Pr\{D > UCL\,|\,\bar{c}\} \\ &= \Pr\{D < 0\,|\,4\} + \left[1 - \Pr\{D \le 10\,|\,4\}\right] \\ &= 0 + \left[1 - POI(10,4)\right] \\ &= 1 - 0.997 \\ &= 0.003 \end{aligned}$$

where $POI(\cdot)$ is the cumulative Poisson distribution.

6-61.   Suppose that we wish to design a control chart for nonconformities per unit with L-sigma limits. Find the minimum sample size that would result in a positive lower control limit for this chart.

c: nonconformities per unit; L: sigma control limits

$$n\bar{c} - L\sqrt{n\bar{c}} > 0$$

$$n\bar{c} > L\sqrt{n\bar{c}}$$

$$n > L^2 / \bar{c}$$

6-65. A paper by R. N. Rodriguez ("Health Care Applications of Statistical Process Control: Examples Using the SAS® System," SAS Users Group International: Proceedings of the 21st Annual Conference, 1996) illustrated several informative applications of control charts to the health care environment. One of these showed how a control chart was employed to analyze the rate of CAT scans performed each month at a clinic. The data used in this example follow.

Month	NSCANB	MMSB	Days	NYRSB
Jan 94	50	26838	31	2.31105
Feb 94	44	26903	28	2.09246
March 94	71	26895	31	2.31596
Apr 94	53	26289	30	2.19075
May 94	53	26149	31	2.25172
Jun 94	40	26185	30	2.18208
Jul 94	41	26142	31	2.25112
Aug 94	57	26092	31	2.24681
Sept 94	49	25958	30	2.16317
Oct 94	63	25957	31	2.23519
Nov 94	64	25920	30	2.16000
Dec 94	62	25907	31	2.23088
Jan 95	67	26754	31	2.30382
Feb 95	58	26696	28	2.07636
Mar 95	89	26565	31	2.28754

NSCANB is the number of CAT scans performed each month and MMSB is the number of members enrolled in the health care plan each month, in units of member months. DAYS is the number of days in each month. The variable NYRSB converts MMSB to units of thousand members per year, and is computed as follows: NYRSB = MMSB(Days/30)/12000. NYRSB represents the "area of opportunity." Construct an appropriate control chart to monitor the rate at which CAT scans are performed at this clinic.

The variable NYRSB can be thought of as an "inspection unit", representing an identical "area of opportunity" for each "sample". The "process characteristic" to be controlled is the rate of CAT scans. A u chart which monitors the average number of CAT scans per NYRSB is appropriate.

MTB>Stat>Control Charts>Attributes Charts>U

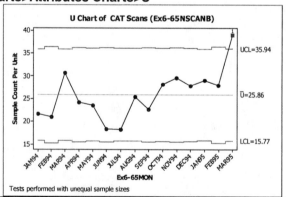

Test Results for U Chart of Ex6-65NSCANB
TEST 1. One point more than 3.00 standard deviations from center line.
Test Failed at points: 15

u chart: UCL = 35.95, CL = 25.86, LCL = 15.77

The rate of monthly CAT scans is out of control.

CHAPTER 7

Process and Measurement System Capability Analysis

Learning Objectives

After completing this chapter you should be able to:
1. Investigate and analyze process capability using control charts, histograms, and probability plots
2. Understand the difference between process capability and process potential
3. Calculate and properly interpret process capability ratios
4. Understand the role of the normal distribution interpreting most process capability ratios
5. Calculate confidence intervals on process capability ratios
6. Know how to conduct and analyze a measurement systems capability (or gauge R&R) experiment
7. Know how to estimate the components of variability in a measurement system
8. Know how to set specifications on components in a system involving interaction components to ensure that overall system requirements are met
9. Estimate the natural limits of a process from a sample of data from that process

Important Terms and Concepts

ANOVA approach to a gauge R&R experiment
Components of measurement error
Confidence intervals on process capability ratios
Control charts and process capability analysis
Discrimination ratio DR for a gauge
Factorial experiment
Graphical methods for process capability analysis
Natural tolerance limits for a normal distribution
Nonparametric tolerance limits
One-sided process capability ratios
PCR Cp
PCR Cpm
Process capability
Process performance indices Pp and Ppk
Product characterization
Signal-to-noise ratio SNR for a gauge
Transmission of error formula

Components of gauge error
Confidence intervals for gauge R&R studies
Consumer's risk or missed fault for a gauge
Delta method
Estimating variance components
Gauge R&R experiment
Measurements systems capability analysis
Natural tolerance limits of a process
Normal distribution and process capability ratios
P/T ratio
PCR Cpk
Precision versus accuracy of a gauge
Process capability analysis
Producer's risk or false failure for a gauge
Random effects model ANOVA
Tolerance stack-up problems

Exercises

7-1. Consider the piston-ring data in Table 5-3. Estimate the process capability assuming that specifications are 74.00 ± 0.035 mm.

Table 5-3   Inside Diameter Measurements (mm) for Automobile Engine Piston Rings

Sample Number	Observations					$\bar{x}_i$	S_i
1	74.030	74.002	74.019	73.992	74.008	74.010	0.0148
2	73.995	73.992	74.001	74.011	74.004	74.001	0.0075
3	73.988	74.024	74.021	74.005	74.002	74.008	0.0147
4	74.002	73.996	73.993	74.015	74.009	74.003	0.0091
5	73.992	74.007	74.015	73.989	74.014	74.003	0.0122
6	74.009	73.994	73.997	73.985	73.993	73.996	0.0087
7	73.995	74.006	73.994	74.000	74.005	74.000	0.0055
8	73.985	74.003	73.993	74.015	73.988	73.997	0.0123
9	74.008	73.995	74.009	74.005	74.004	74.004	0.0055
10	73.998	74.000	73.990	74.007	73.995	73.998	0.0063
11	73.994	73.998	73.994	73.995	73.990	73.994	0.0029
12	74.004	74.000	74.007	74.000	73.996	74.001	0.0042
13	73.983	74.002	73.998	73.997	74.012	73.998	0.0105
14	74.006	73.967	73.994	74.000	73.984	73.990	0.0153
15	74.012	74.014	73.998	73.999	74.007	74.006	0.0073
16	74.000	73.984	74.005	73.998	73.996	73.997	0.0078
17	73.994	74.012	73.986	74.005	74.007	74.001	0.0106
18	74.006	74.010	74.018	74.003	74.000	74.007	0.0070
19	73.984	74.002	74.003	74.005	73.997	73.998	0.0085
20	74.000	74.010	74.013	74.020	74.003	74.009	0.0080
21	73.982	74.001	74.015	74.005	73.996	74.000	0.0122
22	74.004	73.999	73.990	74.006	74.009	74.002	0.0074
23	74.010	73.989	73.990	74.009	74.014	74.002	0.0119
24	74.015	74.008	73.993	74.000	74.010	74.005	0.0087
25	73.982	73.984	73.995	74.017	74.013	73.998	0.0162

$$\Sigma = 1850.028 \qquad 0.2351$$
$$\bar{\bar{x}} = 74.001 \qquad \bar{s} = 0.0094$$

$\hat{\mu} = \bar{\bar{x}} = 74.001; \quad \bar{R} = 0.023; \quad \hat{\sigma} = \bar{R}/d_2 = 0.023/2.326 = 0.010$

$SL = 74.000 \pm 0.035 = [73.965, 74.035]$

$\hat{C}_p \dfrac{USL - LSL}{6\hat{\sigma}} = \dfrac{74.035 - 73.965}{6(0.010)} = 1.17$

$\hat{C}_{pl} = \dfrac{\hat{\mu} - LSL}{3\hat{\sigma}} = \dfrac{74.001 - 73.965}{3(0.010)} = 1.20$

(Equations 7-5, -7, -8, and -9)

$\hat{C}_{pu} = \dfrac{USL - \hat{\mu}}{3\hat{\sigma}} = \dfrac{74.035 - 74.001}{3(0.010)} = 1.13$

$\hat{C}_{pk} = \min\left(\hat{C}_{pl}, \hat{C}_{pu}\right) = 1.13$

7-3. Estimate process capability using $\overline{x}$ and R charts for the power supply voltage data in Exercise 5-2. If specifications are at 350 ± 5 V, calculate Cp, Cpk, and Cpkm. Interpret these capability ratios.

Sample Number	x_1	x_2	x_3	x_4
1	6	9	10	15
2	10	4	6	11
3	7	8	10	5
4	8	9	6	13
5	9	10	7	13
6	12	11	10	10
7	16	10	8	9
8	7	5	10	4
9	9	7	8	12
10	15	16	10	13
11	8	12	14	16
12	6	13	9	11
13	16	9	13	15
14	7	13	10	12
15	11	7	10	16
16	15	10	11	14
17	9	8	12	10
18	15	7	10	11
19	8	6	9	12
20	13	14	11	15

$\hat{\mu} = \overline{\overline{x}} = 10.375; \ \overline{R}_x = 6.25; \ \hat{\sigma}_x = \overline{R}/d_2 = 6.25/2.059 = 3.04$

$\text{USL}_x = [(350+5)-350]\times10 = 50; \ \text{LSL}_x = [(350-5)-350]\times10 = -50$

$x_i = (\text{obs}_i - 350)\times10$

$\hat{C}_p = \dfrac{\text{USL}_x - \text{LSL}_x}{6\hat{\sigma}_x} = \dfrac{50-(-50)}{6(3.04)} = 5.48$ (Equation 7-5)

The process produces product that uses approximately 18% of the total specification band.

$\hat{C}_{pu} = \dfrac{\text{USL}_x - \hat{\mu}}{3\hat{\sigma}_x} = \dfrac{50-10.375}{3(3.04)} = 4.34; \quad \hat{C}_{pl} = \dfrac{\hat{\mu} - \text{LSL}_x}{3\hat{\sigma}_x} = \dfrac{10.375-(-50)}{3(3.04)} = 6.62$ (Eqns 7-7, -8 and -9)

$\hat{C}_{pk} = \min(\hat{C}_{pu}, \hat{C}_{pl}) = 4.34$

This is an extremely capable process, with an estimated percent defective much less than 1 ppb. Note that the C_{pk} is less than C_p, indicating that the process is not centered and is not achieving potential capability. However, this PCR does not tell *where* the mean is located within the specification band.

$V = \dfrac{T - \overline{\overline{x}}}{S} = \dfrac{0-10.375}{3.04} = -3.4128; \quad \hat{C}_{pm} = \dfrac{\hat{C}_p}{\sqrt{1+V^2}} = \dfrac{5.48}{\sqrt{1+(-3.4128)^2}} = 1.54$ (Equations 7-15, -16)

Since C_{pm} is greater than 4/3, the mean μ lies within approximately the middle fourth of the specification band.

$\hat{\xi} = \dfrac{\hat{\mu} - T}{\hat{\sigma}} = \dfrac{10.375-0}{3.04} = 3.41; \quad \hat{C}_{pkm} = \dfrac{\hat{C}_{pk}}{\sqrt{1+\hat{\xi}^2}} = \dfrac{1.54}{\sqrt{1+3.41^2}} = 0.43$ (Equations 7—14, 18)

7-5. A process is in control with $\bar{\bar{x}} = 100$, $\bar{s} = 1.05$, and $n = 5$. The process specifications are at 95 ± 10. The quality characteristic has a normal distribution.

(a) Estimate the potential capability.

$$\hat{\mu} = \bar{\bar{x}} = 100; \bar{s} = 1.05; \hat{\sigma}_x = \bar{s}/c_4 = 1.05/0.9400 = 1.117$$

$$\text{Potential: } \hat{C}_p = \frac{\text{USL} - \text{LSL}}{6\hat{\sigma}} = \frac{(95+10)-(95-10)}{6(1.117)} = 2.98$$

(b) Estimate the actual capability.

$$\hat{C}_{pl} = \frac{\hat{\mu} - \text{LSL}_x}{3\hat{\sigma}_x} = \frac{100-(95-10)}{3(1.117)} = 4.48$$

$$\text{Actual: } \hat{C}_{pu} = \frac{\text{USL}_x - \hat{\mu}}{3\hat{\sigma}_x} = \frac{(95+10)-100}{3(1.117)} = 1.49 \ \text{(Equations 7-7, -8, and -9)}$$

$$\hat{C}_{pk} = \min(\hat{C}_{pl}, \hat{C}_{pu}) = 1.49$$

(c) How much could the fallout in the process be reduced if the process were corrected to operate at the nominal specification?

$$\hat{p}_{\text{Actual}} = \Pr\{x < \text{LSL}\} + \Pr\{x > \text{USL}\}$$

$$= \Pr\{x < \text{LSL}\} + \left[1 - \Pr\{x \le \text{USL}\}\right]$$

$$= \Pr\left\{z < \frac{\text{LSL} - \hat{\mu}}{\hat{\sigma}}\right\} + \left[1 - \Pr\left\{z \le \frac{\text{USL} - \hat{\mu}}{\hat{\sigma}}\right\}\right]$$

$$= \Pr\left\{z < \frac{85 - 100}{1.117}\right\} + \left[1 - \Pr\left\{z \le \frac{105 - 100}{1.117}\right\}\right]$$

$$= \Phi(-13.429) + \left[1 - \Phi(4.476)\right]$$

$$= 0.0000 + \left[1 - 0.999996\right]$$

$$= 0.000004$$

$$\hat{p}_{\text{Potential}} = \Pr\left\{z < \frac{85 - 95}{1.117}\right\} + \left[1 - \Pr\left\{z \le \frac{105 - 95}{1.117}\right\}\right]$$

$$= \Phi(-8.953) + \left[1 - \Phi(8.953)\right]$$

$$= 0.000000 + \left[1 - 1.000000\right]$$

$$= 0.000000$$

7-7.   A process is in statistical control with $\bar{\bar{x}} = 39.7$ and $\bar{R} = 2.5$. The control chart uses a sample size of $n = 2$. Specifications are at 40 ± 5. The quality characteristic is normally distributed.

(a) Estimate the potential capability of the process.

$$n = 2; \quad \hat{\mu} = \bar{\bar{x}} = 39.7; \quad \bar{R} = 2.5; \quad \hat{\sigma}_x = \bar{R}/d_2 = 2.5/1.128 = 2.216$$

$$\text{USL} = 40 + 5 = 45; \text{LSL} = 40 - 5 = 35$$

$$\text{Potential:} \quad \hat{C}_p = \frac{\text{USL} - \text{LSL}}{6\hat{\sigma}} = \frac{45 - 35}{6(2.216)} = 0.75$$

(b) Estimate the actual process capability.

$$\hat{C}_{pu} = \frac{\text{USL} - \hat{\mu}}{3\hat{\sigma}} = \frac{45 - 39.7}{3(2.216)} = 0.80$$

$$\text{Actual:} \quad \hat{C}_{pl} = \frac{\hat{\mu} - \text{LSL}}{3\hat{\sigma}} = \frac{39.7 - 35}{3(2.216)} = 0.71$$

$$\hat{C}_{pk} = \min(\hat{C}_{pl}, \hat{C}_{pu}) = 0.71$$

(c) Calculate and compare the PCRs C_{pm} and C_{pkm}.

$$V = \frac{\bar{x} - T}{s} = \frac{39.7 - 40}{2.216} = -0.135$$

$$\hat{C}_{pm} = \frac{\hat{C}_p}{\sqrt{1 + V^2}} = \frac{0.75}{\sqrt{1 + (-0.135)^2}} = 0.74; \quad \hat{C}_{pkm} = \frac{\hat{C}_{pk}}{\sqrt{1 + V^2}} = \frac{0.71}{\sqrt{1 + (-0.135)^2}} = 0.70$$

The closeness of estimates for C_p, C_{pk}, C_{pm}, and C_{pkm} indicate that the process mean is very close to the specification target.

(d) How much improvement could be made in process performance if the mean could be centered at the nominal value?

The current fraction nonconforming is:

$$\hat{p}_{\text{Actual}} = \Pr\{x < \text{LSL}\} + \Pr\{x > \text{USL}\}$$

$$= \Pr\{x < \text{LSL}\} + \left[1 - \Pr\{x \le \text{USL}\}\right]$$

$$= \Pr\left\{z < \frac{\text{LSL} - \hat{\mu}}{\hat{\sigma}}\right\} + \left[1 - \Pr\left\{z \le \frac{\text{USL} - \hat{\mu}}{\hat{\sigma}}\right\}\right]$$

$$= \Pr\left\{z < \frac{35 - 39.7}{2.216}\right\} + \left[1 - \Pr\left\{z \le \frac{45 - 39.7}{2.216}\right\}\right]$$

$$= \Phi(-2.12094) + \left[1 - \Phi(2.39170)\right] = 0.0169634 + \left[1 - 0.991615\right] = 0.025348$$

If the process mean could be centered at the specification target, the fraction nonconforming would be:

$$\hat{p}_{\text{Potential}} = 2 \times \Pr\left\{z < \frac{35 - 40}{2.216}\right\} = 2 \times \Pr\{z < -2.26\} = 2 \times 0.01191 = 0.02382$$

7-9. Consider the two processes shown here (the sample size $n = 5$):

Process A	Process B
$\bar{x}_A = 100$	$\bar{x}_B = 105$
$\bar{s}_A = 3$	$\bar{s}_B = 1$

Specifications are at 100 ± 10. Calculate C_p, C_{pk}, and C_{pm} and interpret these ratios. Which process would you prefer to use?

Process A

$$\hat{\mu} = \bar{\bar{x}}_A = 100; \bar{s}_A = 3; \hat{\sigma}_A = \bar{s}_A/c_4 = 3/0.9400 = 3.191$$

$$\hat{C}_p = \frac{USL - LSL}{6\hat{\sigma}} = \frac{(100+10) - (100-10)}{6(3.191)} = 1.045$$

$$\hat{C}_{pu} = \frac{USL_x - \hat{\mu}}{3\hat{\sigma}_x} = \frac{(100+10)-100}{3(3.191)} = 1.045; \quad \hat{C}_{pl} = \frac{\hat{\mu} - LSL_x}{3\hat{\sigma}_x} = \frac{100-(100-10)}{3(3.191)} = 1.045$$

$$\hat{C}_{pk} = \min(\hat{C}_{pl}, \hat{C}_{pu}) = 1.045$$

$$V = \frac{\bar{x} - T}{s} = \frac{100-100}{3.191} = 0; \quad \hat{C}_{pm} = \frac{\hat{C}_p}{\sqrt{1+V^2}} = \frac{1.045}{\sqrt{1+(0)^2}} = 1.045$$

$$\hat{p} = \Pr\{x < LSL\} + \Pr\{x > USL\}$$
$$= \Pr\{x < LSL\} + 1 - \Pr\{x \leq USL\}$$
$$= \Pr\left\{z < \frac{LSL - \hat{\mu}}{\hat{\sigma}}\right\} + 1 - \Pr\left\{z \leq \frac{USL - \hat{\mu}}{\hat{\sigma}}\right\}$$
$$= \Pr\left\{z < \frac{90-100}{3.191}\right\} + 1 - \Pr\left\{z \leq \frac{110-100}{3.191}\right\}$$
$$= \Phi(-3.13) + 1 - \Phi(3.13) = 0.00087 + 1 - 0.99913 = 0.00174$$

Process B

$$\hat{\mu} = \bar{\bar{x}}_B = 105; \bar{s}_B = 1; \hat{\sigma}_B = \bar{s}_B/c_4 = 1/0.9400 = 1.064$$

$$\hat{C}_p = \frac{USL - LSL}{6\hat{\sigma}} = \frac{(100+10) - (100-10)}{6(1.064)} = 3.133$$

$$\hat{C}_{pl} = \frac{\hat{\mu}_x - LSL_x}{3\hat{\sigma}_x} = \frac{105 - (100-10)}{3(1.064)} = 4.699; \quad \hat{C}_{pu} = \frac{USL_x - \hat{\mu}_x}{3\hat{\sigma}_x} = \frac{(100+10)-105}{3(1.064)} = 1.566$$

$$\hat{C}_{pk} = \min(\hat{C}_{pl}, \hat{C}_{pu}) = 1.566$$

$$V = \frac{\bar{x} - T}{s} = \frac{100-105}{1.064} = -4.699; \quad \hat{C}_{pm} = \frac{\hat{C}_p}{\sqrt{1+V^2}} = \frac{3.133}{\sqrt{1+(-4.699)^2}} = 0.652$$

$$\hat{p} = \Pr\left\{z < \frac{90-105}{1.064}\right\} + 1 - \Pr\left\{z \leq \frac{110-105}{1.064}\right\}$$
$$= \Phi(-14.098) + 1 - \Phi(4.699) = 0.000000 + 1 - 0.999999 = 0.000001$$

Prefer to use Process B with estimated process fallout of 0.000001 instead of Process A with estimated fallout 0.001726.

7-11. The weights of nominal 1-kg containers of a concentrated chemical ingredient are shown here. Prepare a normal probability plot of the data and estimate process capability.

0.9475	0.9775	0.9965	1.0075	1.0180
0.9705	0.9860	0.9975	1.0100	1.0200
0.9770	0.9960	1.0050	1.0175	1.0250

MTB>Stat>Basic Statistics>Normality Test

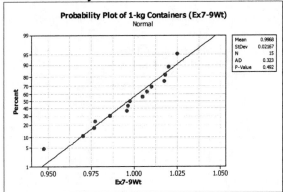

A normal probability plot of the 1-kg container weights shows the distribution is close to normal.

$$\overline{x} \approx p_{50} = 0.9975; \quad p_{84} = 1.0200$$
$$\hat{\sigma} = p_{84} - p_{50} = 1.0200 - 0.9975 = 0.0225$$
$$6\hat{\sigma} = 6(0.0225) = 0.1350$$

7-13. The height of the disk used in a computer disk drive assembly is a critical quality characteristic. Below are the heights (in mm) of 25 disks randomly selected from the manufacturing process.

20.0106	20.0090	20.0067	19.9772	20.0001
19.9940	19.9876	20.0042	19.9986	19.9958
20.0075	20.0018	20.0059	19.9975	20.0089
20.0045	19.9891	19.9956	19.9884	20.0154
20.0056	19.9831	20.0040	20.0006	20.0047

Prepare a normal probability plot of the disk height data and estimate process capability.

MTB>Stat>Basic Statistics>Normality Test
(Add percentile lines at Y values 50 and 84 to estimate μ and σ.)

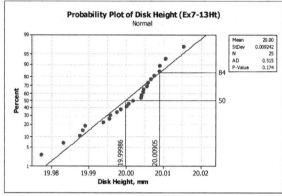

A normal probability plot of computer disk heights shows the distribution is close to normal.

$$\bar{x} \approx p_{50} = 19.99986$$

$$p_{84} = 20.00905$$

$$\hat{\sigma} = p_{84} - p_{50} = 20.00905 - 19.99986 = 0.00919$$

$$6\hat{\sigma} = 6(0.00919) = 0.05514$$

7-15. An electric utility tracks the response time to customer-reported outages. The following data are a random sample of 40 of the response times (in minutes) for one operating division of this utility during a single month.

80	102	86	94	86	106	105	110	127	97
110	104	97	128	98	84	97	87	99	94
105	104	84	77	125	85	80	104	106	109
115	89	100	96	96	87	106	100	102	93

MTB>Stat>Basic Statistics>Normality Test
(Add percentile lines at Y values 50 and 84 to estimate μ and σ.)

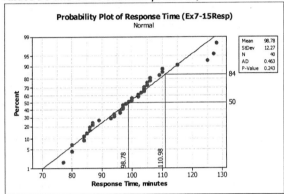

A normal probability plot of response times shows the distribution is close to normal.

(a) Estimate the capability of the utility's process for responding to customer-reported outages.

$\bar{x} \approx p_{50} = 98.78$

$p_{84} = 110.98$

$\hat{\sigma} = p_{84} - p_{50} = 110.98 - 98.78 = 12.2$

$6\hat{\sigma} = 6(12.2) = 73.2$

(b) The utility wants to achieve a 90% response rate in under two hours, as response to emergency outages is an important measure of customer satisfaction. What is the capability of the process with respect to this objective?

USL = 2 hrs = 120 mins

$C_{pu} = \dfrac{\text{USL} - \hat{\mu}}{3\hat{\sigma}} = \dfrac{120 - 98.78}{3(12.2)} = 0.58$

$\hat{p} = \Pr\left\{z > \dfrac{\text{USL} - \hat{\mu}}{\hat{\sigma}}\right\} = 1 - \Pr\left\{z < \dfrac{\text{USL} - \hat{\mu}}{\hat{\sigma}}\right\} = 1 - \Pr\left\{z < \dfrac{120 - 98.78}{12.2}\right\}$

$= 1 - \Phi(1.739) = 1 - 0.958983 = 0.041017$

7-17.   The failure time in hours of 10 LSI memory devices follows: 1210, 1275, 1400, 1695, 1900, 2105, 2230, 2250, 2500, and 2625. Plot the data on normal probability paper and, if appropriate, estimate process capability. Is it safe to estimate the proportion of circuits that fail below 1200 h?

MTB > Stat > Basic Statistics > Normality Test

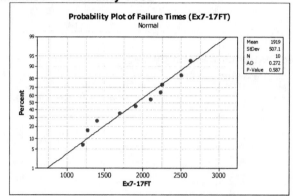

The plot shows that the data is not normally distributed; so it is not appropriate to estimate capability.

7-19.   A company has been asked by an important customer to demonstrate that its process capability ratio C_p exceeds 1.33. It has taken a sample of 50 parts and obtained the point estimate $\hat{C}_p = 1.52$. Assume that the quality characteristic follows a normal distribution. Can the company demonstrate that C_p exceeds 1.33 at the 95% level of confidence? What level of confidence would give a one-sided lower confidence limit on C_p that exceeds 1.33?

$$n = 50$$

$$\hat{C}_p = 1.52$$

$$1 - \alpha = 0.95$$

$$\chi^2_{1-\alpha,n-1} = \chi^2_{0.95,49} = 33.9303$$

$$\hat{C}_p \sqrt{\frac{\chi^2_{1-\alpha,n-1}}{n-1}} \leq C_p$$

$$1.52\sqrt{\frac{33.9303}{49}} = 1.26 \leq C_p$$

The company cannot demonstrate that the PCR exceeds 1.33 at a 95% confidence level.

$$1.52\sqrt{\frac{\chi^2_{1-\alpha,49}}{49}} = 1.33$$

$$\chi^2_{1-\alpha,49} = 49\left(\frac{1.33}{1.52}\right)^2 = 37.52$$

$$1 - \alpha = 0.88$$

$$\alpha = 0.12$$

7-21.   The molecular weight of a particular polymer should fall between 2100 and 2350. Fifty samples of this material were analyzed with the results $\bar{x} = 2275$ and $s = 60$. Assume that molecular weight is normally distributed.

(a) Calculate a point estimate of C_{pk}.

USL = 2350; LSL = 2100; nominal = 2225; $\bar{x} = 2275$; $s = 60$; $n = 50$

$$\hat{C}_{pu} = \frac{\text{USL}_x - \hat{\mu}_x}{3\hat{\sigma}_x} = \frac{2350 - 2275}{3(60)} = 0.42; \quad \hat{C}_{pl} = \frac{\hat{\mu}_x - \text{LSL}_x}{3\hat{\sigma}_x} = \frac{2275 - 2100}{3(60)} = 0.97$$

$$\hat{C}_{pk} = \min(\hat{C}_{pl}, \hat{C}_{pu}) = 0.42$$

(b) Find a 95% confidence interval on C_{pk}.

$\alpha = 0.05$; $z_{\alpha/2} = z_{0.025} = 1.960$

$$\hat{C}_{pk}\left[1 - z_{\alpha/2}\sqrt{\frac{1}{9n\hat{C}_{pk}^2} + \frac{1}{2(n-1)}}\right] \le C_{pk} \le \hat{C}_{pk}\left[1 + z_{\alpha/2}\sqrt{\frac{1}{9n\hat{C}_{pk}^2} + \frac{1}{2(n-1)}}\right]$$

$$0.42\left[1 - 1.96\sqrt{\frac{1}{9(50)(0.42)^2} + \frac{1}{2(50-1)}}\right] \le C_{pk} \le 0.42\left[1 + 1.96\sqrt{\frac{1}{9(50)(0.42)^2} + \frac{1}{2(50-1)}}\right]$$

$$0.2957 \le C_{pk} \le 0.5443$$

7-23.   An operator–instrument combination is known to test parts with an average error of zero; however, the standard deviation of measurement error is estimated to be 3. Samples from a controlled process were analyzed, and the total variability was estimated to be $\hat{\sigma} = 5$. What is the true process standard deviation?

$$\sigma_{OI} = 0; \; \hat{\sigma}_I = 3; \; \hat{\sigma}_{\text{Total}} = 5$$

$$\hat{\sigma}_{\text{Total}}^2 = \hat{\sigma}_{\text{Meas}}^2 + \hat{\sigma}_{\text{Process}}^2$$

$$\hat{\sigma}_{\text{Process}} = \sqrt{\hat{\sigma}_{\text{Total}}^2 - \hat{\sigma}_{\text{Meas}}^2} = \sqrt{5^2 - 3^2} = 4$$

7-25. Ten parts are measured three times by the same operator in a gauge capability study. The data are shown here.

Part	Measurements		
Number	1	2	3
1	100	101	100
2	95	93	97
3	101	103	100
4	96	95	97
5	98	98	96
6	99	98	98
7	95	97	98
8	100	99	98
9	100	100	97
10	100	98	99

(a) Describe the measurement error that results from the use of this gauge.

MTB>Stat>Control Charts>Variables Charts for Subgroups>X-bar R

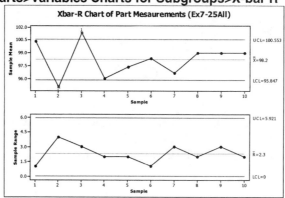

Test Results for Xbar Chart of Ex7-25All
```
TEST 1. One point more than 3.00 standard deviations from center line.
Test Failed at points:  2, 3
```

$\bar{x}$ chart: UCL = 100.553, CL = 98.2, LCL = 95.847
R chart: UCL = 5.921, CL = 2.3, LCL = 0

The $\bar{x}$ chart has a couple out-of-control points, and the R chart is in control. This indicates that the operator is not having difficulty making consistent measurements.

(b) Estimate total variability and product variability.

$$\bar{\bar{x}} = 98.2; \bar{R} = 2.3; \hat{\sigma}_{Gauge} = \bar{R}/d_2 = 2.3/1.693 = 1.359$$

$$\hat{\sigma}^2_{Total} = 4.717$$

$$\hat{\sigma}^2_{Product} = \hat{\sigma}^2_{Total} - \hat{\sigma}^2_{Gauge} = 4.717 - 1.359^2 = 2.872$$

$$\hat{\sigma}_{Product} = 1.695$$

7-25 continued

(c) What percentage of total variability is due to the gauge?

$$\frac{\hat{\sigma}_{Gauge}}{\hat{\sigma}_{Total}} \times 100 = \frac{1.359}{\sqrt{4.717}} \times 100 = 62.5\%$$

(d) If specifications on the part are at 100 ± 15, find the P/T ratio for this gauge. Comment on the adequacy of the gauge.

$$USL = 100 + 15 = 115; LSL = 100 - 15 = 85$$

$$\frac{P}{T} = \frac{6\hat{\sigma}_{Gauge}}{USL - LSL} = \frac{6(1.359)}{115 - 85} = 0.272$$

7-27. The following data were taken by one operator during a gauge capability study.

Part Number	Measurements 1	2	Part Number	Measurements 1	2
1	20	20	9	20	20
2	19	20	10	23	22
3	21	21	11	28	22
4	24	20	12	19	25
5	21	21	13	21	20
6	25	26	14	20	21
7	18	17	15	18	18
8	16	15			

(a) Estimate gauge capability.

$$\hat{\sigma}_{Gauge} = \overline{R}/d_2 = 1.533/1.128 = 1.359$$

Gauge capability: $6\hat{\sigma} = 8.154$

(b) Does the control chart analysis of these data indicate any potential problem in using the gauge?

MTB>Stat>Control Charts>Variables Charts for Subgroups>X-bar R

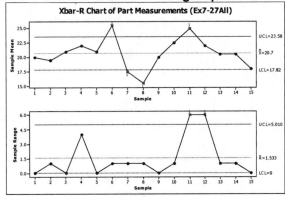

Xbar-R Chart of Part Measurements (Ex7-27All)

Test Results for R Chart of Ex7-27All
```
TEST 1. One point more than 3.00 standard deviations from center line.
Test Failed at points:  11, 12
```

Out-of-control points on R chart indicate operator difficulty with using gage.

7-29. Reconsider the gauge R & R experiment in Exercise 7-28. Calculate the quantities *SNR* and *DR* for this gauge. Discuss what information these measures provide about the capability of the gauge.

$$\hat{\sigma}^2_{Part} = 10.2513;\ \hat{\sigma}^2_{Total} = 11.1451 \quad \text{(from Exercise 5-28)}$$

$$\hat{\rho}_P = \frac{\hat{\sigma}^2_{Part}}{\hat{\sigma}^2_{Total}} = \frac{10.2513}{11.1451} = 0.9198$$

$$SNR = \sqrt{\frac{2\hat{\rho}_P}{1-\hat{\rho}_P}} = \sqrt{\frac{2(0.9198)}{1-0.9198}} = 4.79 \qquad \text{(Equations 7-26, -28, -29)}$$

$$DR = \frac{1+\hat{\rho}_P}{1-\hat{\rho}_P} = \frac{1+0.9198}{1-0.9198} = 23.94$$

SNR = 4.79 indicates that fewer than five distinct levels can be reliably obtained from the measurements. This is near the AIAG-recommended value of five levels or more, but larger than a value of two (or less) that indicates inadequate gauge capability. (Also note that the MINITAB Gage R&R output indicates "Number of Distinct Categories = 4"; this is also the number of distinct categories of parts that the gauge is able to distinguish)

DR = 23.94, exceeding the minimum recommendation of four. By this measure, the gauge is capable.

7-31. Two parts are assembled as shown in the figure. The distributions of x_1 and x_2 are normal, with $\mu_1 = 20$, $\sigma_1 = 0.3$, $\mu_2 = 19.6$, and $\sigma_2 = 0.4$. The specifications of the clearance between the mating parts are 0.5 ± 0.4. What fraction of assemblies will fail to meet specifications if assembly is at random?

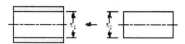

$x_1 \sim N(20, 0.3^2);\ x_2 \sim N(19.6, 0.4^2)$

Nonconformities will occur if $y = x_1 - x_2 < 0.1$ or $y = x_1 - x_2 > 0.9$

$\mu_y = \mu_1 - \mu_2 = 20 - 19.6 = 0.4$

$\sigma^2_y = \sigma^2_1 + \sigma^2_2 = 0.3^2 + 0.4^2 = 0.25$

$\sigma_y = 0.50$

$$\begin{aligned}
Pr\{Nonconformities\} &= Pr\{y < LSL\} + Pr\{y > USL\} \\
&= Pr\{y < 0.1\} + Pr\{y > 0.9\} \\
&= Pr\{y < 0.1\} + 1 - Pr\{y \le 0.9\} \\
&= \Phi\left(\frac{0.1-0.4}{\sqrt{0.25}}\right) + 1 - \Phi\left(\frac{0.9-0.4}{\sqrt{0.25}}\right) \\
&= \Phi(-0.6) + 1 - \Phi(1.00) \\
&= 0.2743 + 1 - 0.8413 \\
&= 0.4330
\end{aligned}$$

7-33.   A rectangular piece of metal of width W and length L is cut from a plate of thickness T. If W, L, and T are independent random variables with means and standard deviations as given here and the density of the metal is 0.08 g/cm^3, what would be the estimated mean and standard deviation of the weights of pieces produced by this process?

$$\text{Weight} = d \times W \times L \times T$$
$$\cong d\left[\mu_W \mu_L \mu_T + (W - \mu_W)\mu_L \mu_T + (L - \mu_L)\mu_W \mu_T + (T - \mu_T)\mu_W \mu_L\right]$$
$$\hat{\mu}_{\text{Weight}} \cong d[\mu_W \mu_L \mu_T] = 0.08(10)(20)(3) = 48$$
$$\hat{\sigma}^2_{\text{Weight}} \cong d^2\left[\hat{\mu}^2_W \hat{\sigma}^2_L \hat{\sigma}^2_T + \hat{\mu}^2_L \hat{\sigma}^2_W \hat{\sigma}^2_T + \hat{\mu}^2_T \hat{\sigma}^2_W \hat{\sigma}^2_L\right]$$
$$= 0.08^2\left[10^2(0.3^2)(0.1^2) + 20^2(0.2^2)(0.1^2) + 3^2(0.2^2)(0.3^2)\right] = 0.00181$$
$$\hat{\sigma}_{\text{Weight}} \cong 0.04252$$

7-35.   Two resistors are connected to a battery as shown in the figure. Find approximate expressions for the mean and variance of the resulting current (I). E, R_1, and R_2 are random variables with means μ_E, μ_{R1}, μ_{R2}, and variances σ^2_E, σ^2_{R1}, and σ^2_{R2}, respectively.

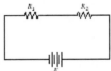

$$I = E/(R_1 + R_2)$$
$$\mu_I \cong \mu_E/(\mu_{R_1} + \mu_{R_2})$$
$$\sigma^2_I \cong \frac{\sigma^2_E}{(\mu_{R_1} + \mu_{R_2})} + \frac{\mu_E}{(\mu_{R_1} + \mu_{R_2})^2}\left(\sigma^2_{R_1} + \sigma^2_{R_2}\right)$$

7-37.   An assembly of two parts is formed by fitting a shaft into a bearing. It is known that the inside diameters of bearings are normally distributed with mean 2.010 cm and standard deviation 0.002 cm, and that the outside diameters of the shafts are normally distributed with mean 2.004 cm and standard deviation 0.001 cm. Determine the distribution of clearance between the parts if random assembly is used. What is the probability that the clearance is positive?

$$\text{ID} \sim N(2.010, 0.002^2) \text{ and OD} \sim N(2.004, 0.001^2)$$
Interference occurs if $y = \text{ID} - \text{OD} < 0$
$$\mu_y = \mu_{\text{ID}} - \mu_{\text{OD}} = 2.010 - 2.004 = 0.006$$
$$\sigma^2_y = \sigma^2_{\text{ID}} + \sigma^2_{\text{OD}} = 0.002^2 + 0.001^2 = 0.000005; \quad \sigma_y = 0.002236$$
$$\Pr\{\text{positive clearance}\} = 1 - \Pr\{\text{interference}\}$$
$$= 1 - \Pr\{y < 0\}$$
$$= 1 - \Phi\left(\frac{0 - 0.006}{\sqrt{0.000005}}\right) = 1 - \Phi(-2.683) = 1 - 0.0036 = 0.9964$$

7-39.   A sample of 10 items from a normal population had a mean of 300 and a standard deviation of 10. Using these data, estimate a value for the random variable such that the probability is 0.95 that 90% of the measurements on this random variable will lie below the value.

$n = 10; \ x \sim N(300,10^2); \ \alpha = 0.10; \ \gamma = 0.95$; one-sided

From Appendix VIII: $K = 2.355$

$\text{UTL} = \bar{x} + KS = 300 + 2.355(10) = 323.55$

7-41.   A sample of 20 measurements on a normally distributed quality characteristic had $\bar{\bar{x}} = 350$ and $s = 10$. Find an upper natural tolerance limit that has probability 0.90 of containing 95% of the distribution of this quality characteristic.

$n = 20; \ x \sim N(350,10^2); \ \alpha = 0.05; \ \gamma = 0.90$; one-sided

From Appendix VIII: $K = 2.208$

$\text{UTL} = \bar{x} + KS = 350 + 2.208(10) = 372.08$

7-43.   A random sample of $n = 40$ pipe sections resulted in a mean wall thickness of 0.1264 in. and a standard deviation of 0.0003 in. We assume that wall thickness is normally distributed.

(a) Between what limits can we say with 95% confidence that 95% of the wall thicknesses should fall?

$x \sim N\left(0.1264, 0.0003^2\right)$

$\alpha = 0.05; \ \gamma = 0.95$; and two-sided

From Appendix VII: $K = 2.445$

TI on $x : \bar{x} \pm KS = 0.1264 \pm 2.445(0.0003) = [0.1257, 0.1271]$

(b) Construct a 95% confidence interval on the true mean thickness. Explain the difference between this interval and the one constructed in part (a).

$\alpha = 0.05; \ t_{\alpha/2,n-1} = t_{0.025,39} = 2.023$

CI on $\bar{x} : \bar{x} \pm t_{\alpha/2,n-1} \, S/\sqrt{n} = 0.1264 \pm 2.023\left(0.0003/\sqrt{40}\right) = [0.1263, 0.1265]$

Part (a) is a tolerance interval on individual thickness observations; part (b) is a confidence interval on mean thickness. In part (a), the interval relates to individual observations (random variables), while in part (b) the interval refers to a parameter of a distribution (an unknown constant).

CHAPTER 8

Cumulative Sum and Exponentially Weighted Moving Average Control Charts

Learning Objectives

After completing this chapter you should be able to:
1. Set up and use cusum control charts for monitoring the process mean
2. Design a cusum control chart for the mean to obtain specific ARL performance
3. Incorporate a fast initial response feature into the cusum control chart
4. Use a combined Shewhart-cusum monitoring scheme
5. Set up and use EWMA control charts for monitoring the process mean
6. Design an EWMA control chart for the mean to obtain specific ARL performance
7. Understand why the EWMA control chart is robust to the assumption of normality
8. Understand the performance advantage of cusum and EWMA control charts relative to Shewhart control charts
9. Set up and use a control chart based on ordinary (unweighted) moving average

Important Terms and Concepts

An EWMA for Poisson data
Average run length
Cusum control chart
Decision interval
Design of an EWMA control chart
Fast initial response (headstart) feature for a cusum
Moving average control chart
Reference value
Scale cusum
Tabular or algorithmic cusum

ARL calculations for the cusum
Combined cusum-Shewhart monitoring schemes
Cusum status chart
Design of a cusum
EWMA control chart
Fast initial response (headstart) feature for an EWMA
One-sided cusums
Robustness of the EWMA to normality
Standardized cusum
V-mask form of the cusum

Exercises

8-1.   The following data represent individual observations on molecular weight taken hourly from a chemical process.

Observation Number	x	Observation Number	x
1	1045	11	1139
2	1055	12	1169
3	1037	13	1151
4	1064	14	1128
5	1095	15	1238
6	1008	16	1125
7	1050	17	1163
8	1087	18	1188
9	1125	19	1146
10	1146	20	1167

The target value of molecular weight is 1050 and the process standard deviation is thought to be about $\sigma = 25$.

(a) Set up a tabular cusum for the mean of this process. Design the cusum to quickly detect a shift of about 1.0 σ in the process mean.

$\mu_0 = 1050$; $\sigma = 25$; $\delta = 1\sigma$, $K = (\delta/2)\sigma = (1/2)25 = 12.5$; $H = 5\sigma = 5(25) = 125$

MTB>Stat>Control Charts>Time-Weighted Charts>CUSUM

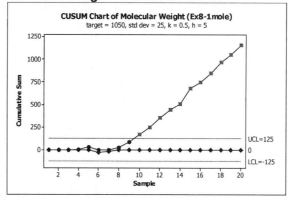

UCL = 125, LCL = - 125

The process signals out of control at observation 10. The point at which the assignable cause occurred can be determined by counting the number of increasing plot points. The assignable cause occurred after observation 10 – 3 = 7.

(b) Is the estimate of σ used in part (a) of this problem reasonable?

$\hat{\sigma} = \overline{MR2}/d_2 = 38.8421/1.128 = 34.4345$ (Equation 5-6)

No. The estimate used for σ is much smaller than that from the data.

8-3.

(a) Add a headstart feature to the cusum in Exercise 8-1.

$\mu_0 = 1050$, $\sigma = 25$, $k = 0.5$, $K = 12.5$, $h = 5$, $H/2 = 125/2 = 62.5$

FIR = $H/2 = 62.5$, in std dev units = $62.5/25 = 2.5$ (text page 398)

MTB>Stat>Control Charts>Time-Weighted Charts>CUSUM

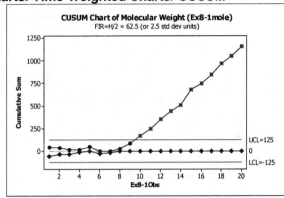

cusum chart: UCL = +125, LCL = -125

For example,

$$C_1^+ = \max\left[0, x_i - (\mu_0 - K) + C_0^+\right] = \max\left[0, 1045 - (1050 + 12.5) + 62.5\right] = 45 \quad \text{(Equation 8-2)}$$

Using the tabular CUSUM, the process signals out of control at observation 10, the same as the CUSUM without a FIR feature.

(b) Use a combined Shewhart-cusum scheme on the data in Exercise 8-1. Interpret the results of both charts.

MTB>Stat>Control Charts>Variables Charts for Individuals>I-MR

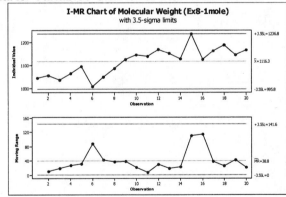

x chart: UCL = 1236.8, CL = 1116.3, LCL = 995.8
MR chart: UCL = 141.6, CL = 38.8, LCL = 0

Using 3.5σ limits on the Individuals chart, there are no out-of-control signals. However there does appear to be a trend up from observations 6 through 12—this is the situation detected by the cumulative sum.

8-5. Rework Exercise 8-4 using the standardized cusum parameters of $h = 8.01$ and $k = 0.25$. Compare the results with those obtained previously in Exercise 8-4. What can you say about the theoretical performance of those two cusum schemes?

$\mu_0 = 8.02$, $\sigma = 0.05$, $k = 0.25$, $h = 8.01$, $H = h\sigma = 8.01\ (0.05) = 0.4005$

MTB>Stat>Control Charts>Time-Weighted Charts>CUSUM

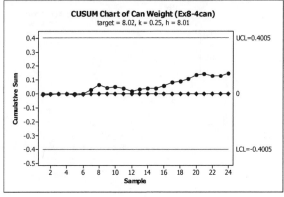

cusum chart: UCL = 0.4005, CL = 0, LCL = -0.4005

There are no out-of-control signals.

<u>In Exercise 8-5:</u>

$\mu_0 = 8.02$; $\sigma = 0.05$; $k = 0.25$; $h = 8.01$; $b = h + 1.166 = 8.01 + 1.166 = 9.176$

$\delta^* = 0$; $\Delta^+ = \delta^* - k = 0 - 0.25 = -0.25$; $\Delta^- = -\delta^* - k = -0 - 0.25 = -0.25$

$ARL_0^+ = ARL_0^- \cong \dfrac{\exp[-2(-0.25)(9.176)] + 2(-0.25)(9.176) - 1}{2(-0.25)^2} = 741.6771$

$\dfrac{1}{ARL_0} = \dfrac{1}{ARL_0^+} + \dfrac{1}{ARL_0^-} = \dfrac{2}{741.6771} = 0.0027$ (Equations 8-6, -7)

$ARL_0 = 1/0.0027 = 370.84$

The theoretical performance of these two CUSUM schemes is the same for Exercises 8-4 and 8-5.

8-7.   The data that follow are temperature readings from a chemical process in °C, taken every 2 minutes. (Read the observations down, from left.) The target value for the mean is $\mu_0 = 950$.

953	985	949	937	959	948	958	952
945	973	941	946	939	937	955	931
972	955	966	954	948	955	947	928
945	950	966	935	958	927	941	937
975	948	934	941	963	940	938	950
970	957	937	933	973	962	945	970
959	940	946	960	949	963	963	933
973	933	952	968	942	943	967	960
940	965	935	959	965	950	969	934
936	973	941	956	962	938	981	927

(a) Estimate the process standard deviation.

Construct a Moving Range chart to find $\overline{MR2}$ and estimate standard deviation.
MTB>Stat>Control Charts>Variables Charts for Individuals>Moving Range

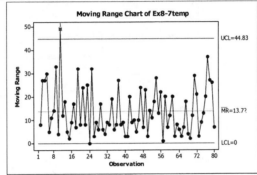

$$\overline{MR2} = 13.72; \quad \hat{\sigma} = \overline{MR2}/d_2 = 13.72/1.128 = 12.16$$

(b) Set up and apply a tabular cusum for this process, using standardized values $h = 5$ and $k = 1\text{-}2$. Interpret this chart.

$$\mu_0 = 950; \hat{\sigma} = 12.16; k = 1/2; h = 5$$

MTB>Stat>Control Charts>Time-Weighted Charts>CUSUM

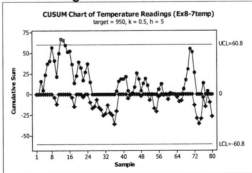

Test Results for CUSUM Chart of Ex8-7temp
```
TEST. One point beyond control limits.
Test Failed at points:  12, 13
```

cusum chart: UCL = +60.8, CL = 0, LCL = -60.8
The process signals out of control at sample 12. The assignable cause occurred after sample 12 – 10 = 2.

8-9. Viscosity measurements on a polymer are made every 10 minutes by an on-line viscometer. Thirty-six observations are shown here (read down from left). The target viscosity for this process is $\mu_0 = 3200$.

3169	3205	3185	3188
3173	3203	3187	3183
3162	3209	3192	3175
3154	3208	3199	3174
3139	3211	3197	3171
3145	3214	3193	3180
3160	3215	3190	3179
3172	3209	3183	3175
3175	3203	3197	3174

(a) Estimate the process standard deviation.

A Moving Range chart has CL = 6.71; $\hat{\sigma} = \overline{MR2}/d_2 = 6.71/1.128 = 5.949$

(b) Construct a tabular cusum for this process using standardized values of $h = 8.01$ and $k = 0.25$.

$\mu_0 = 3200$; $\hat{\sigma} = 5.949$; $k = 0.25$; $h = 8.01$

MTB>Stat>Control Charts>Time-Weighted Charts>CUSUM

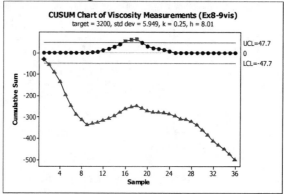

CUSUM Chart of Viscosity Measurements (Ex8-9vis)
target = 3200, std dev = 5.949, k = 0.25, h = 8.01

Test Results for CUSUM Chart of Ex8-9vis
```
TEST. One point beyond control limits.
Test Failed at points:  16, 17, 18
```

cusum chart: UCL = +47.7, CL = 0, LCL = -47.7

The process signals out of control on the lower side at sample 2 and on the upper side at sample 16. Assignable causes occurred after startup (sample 2 – 2) and after sample 9 (16 – 7).

(c) Discuss the choice of h and k in part (b) of this problem on cusum performance.

Selecting a smaller shift to detect, $k = 0.25$, should be balanced by a larger control limit, $h = 8.01$, to give longer in-control ARLs with shorter out-of-control ARLs.

8-11. Apply the scale cusum discussed in Section 8-1.8 to the data in Exercise 8-1.

$$V_i = \left(\sqrt{|y_i|} - 0.822\right)\big/0.349$$

mu0 =	1050		
sigma =	25	(from Exercise 8-1)	
delta =	1 sigma		
k =	0.5		
h =	5		
Obs, i	xi	yi	vi
No FIR			
1	1045	=(B10-$B$1)/$B$2	=(SQRT(ABS(C10))-0.822)/0.349
2	1055	=(B11-$B$1)/$B$2	=(SQRT(ABS(C11))-0.822)/0.349
3	1037	=(B12-$B$1)/$B$2	=(SQRT(ABS(C12))-0.822)/0.349

one-sided upper cusum			
Si+	N+	OOC?	When?
0			
=MAX(0,D10-$B$4+E9)	=IF(E10=0,0,F9+1)	=IF(E10>$B$5,"OOC","")	=IF(G10="OOC",$A10-F10,"")
=MAX(0,D11-$B$4+E10)	=IF(E11=0,0,F10+1)	=IF(E11>$B$5,"OOC","")	=IF(G11="OOC",$A11-F11,"")
=MAX(0,D12-$B$4+E11)	=IF(E12=0,0,F11+1)	=IF(E12>$B$5,"OOC","")	=IF(G12="OOC",$A12-F12,"")

one-sided lower cusum			
Si-	N-	OOC?	When?
0			
=MAX(0,-$B$4-D10+J9)	=IF(J10=0,0,K9+1)	=IF(J10<-$B$5,"OOC","")	=IF(L10="OOC",$A10-K10,"")
=MAX(0,-$B$4-D11+J10)	=IF(J11=0,0,K10+1)	=IF(J11<-$B$5,"OOC","")	=IF(L11="OOC",$A11-K11,"")
=MAX(0,-$B$4-D12+J11)	=IF(J12=0,0,K11+1)	=IF(J12<-$B$5,"OOC","")	=IF(L12="OOC",$A12-K12,"")

mu0 =	1050										
sigma =	25										
delta =	1 sigma										
k =	0.5										
h =	5										
				one-sided upper cusum				one-sided lower cusum			
Obs, i	xi	yi	vi	Si+	N+	OOC?	When?	Si-	N-	OOC?	When?
No FIR				0				0			
1	1045	-0.2	-1.07	0	0			0.57	1		
2	1055	0.2	-1.07	0	0			1.15	2		
3	1037	-0.52	-0.29	0	0			0.94	3		
4	1064	0.56	-0.21	0	0			0.65	4		
5	1095	1.8	1.49	0.989	1			0	0		
6	1008	-1.68	1.36	1.848	2			0	0		
7	1050	0	-2.36	0	0			1.86	1		
8	1087	1.48	1.13	0.631	1			0.22	2		
9	1125	3	2.61	2.738	2			0	0		
10	1146	3.84	3.26	5.498	3	OOC	7	0	0		
11	1139	3.56	3.05	8.049	4	OOC	7	0	0		
12	1169	4.76	3.90	11.44	5	OOC	7	0	0		
13	1151	4.04	3.40	14.35	6	OOC	7	0	0		
14	1128	3.12	2.71	16.55	7	OOC	7	0	0		
15	1238	7.52	5.50	21.56	8	OOC	7	0	0		
16	1125	3	2.61	23.66	9	OOC	7	0	0		
17	1163	4.52	3.74	26.9	10	OOC	7	0	0		
18	1188	5.52	4.38	30.78	11	OOC	7	0	0		
19	1146	3.84	3.26	33.54	12	OOC	7	0	0		
20	1167	4.68	3.84	36.88	13	OOC	7	0	0		

The process is out of control after observation $10 - 3 = 7$. Process variability is increasing.

8-13.   Consider a standardized two-sided cusum with $k = 0.2$ and $h = 8$. Use Siegmund's procedure to evaluate the in-control ARL performance of this scheme. Find ARL_1 for $\delta^* = 0.5$.

Standardized, two-sided cusum with $k = 0.2$ and $h = 8$

<u>In-control ARL performance</u>
$\delta^* = 0$

$\Delta^+ = \delta^* - k = 0 - 0.2 = -0.2$

$\Delta^- = -\delta^* - k = -0 - 0.2 = -0.2$

$b = h + 1.166 = 8 + 1.166 = 9.166$

$ARL_0^+ = ARL_0^- \cong \dfrac{\exp[-2(-0.2)(9.166)] + 2(-0.2)(9.166) - 1}{2(-0.2)^2} = 430.556$

$\dfrac{1}{ARL_0} = \dfrac{1}{ARL_0^+} + \dfrac{1}{ARL_0^-} = \dfrac{2}{430.556} = 0.005$ (Equations 8-6, -7)

$ARL_0 = 1 / 0.005 = 215.23$

<u>Out-of-control ARL Performance</u>
$\delta^* = 0.5$

$\Delta^+ = \delta^* - k = 0.5 - 0.2 = 0.3$

$\Delta^- = -\delta^* - k = -0.5 - 0.2 = -0.7$

$b = h + 1.166 = 8 + 1.166 = 9.166$

$ARL_1^+ = \dfrac{\exp[-2(0.3)(9.166)] + 2(0.3)(9.166) - 1}{2(0.3)^2} = 25.023$

$ARL_1^- = \dfrac{\exp[-2(-0.7)(9.166)] + 2(-0.7)(9.166) - 1}{2(-0.7)^2} = 381,767$ (Equations 8-6, -7)

$\dfrac{1}{ARL_1} = \dfrac{1}{ARL_1^+} + \dfrac{1}{ARL_1^-} = \dfrac{1}{25.023} + \dfrac{1}{381,767} = 0.040$

$ARL_1 = 1 / 0.040 = 25.02$

8-15. Consider the velocity of light data introduced in Exercises 5-59 and 5-60. Use only the 20 observations in Exercise 5-59 to set up a cusum with target value 734.5. Plot all 40 observations from both Exercises 5-59 and 5-60 on this cusum. What conclusions can you draw?

Measurement	Velocity	Measurement	Velocity	Measurement	Velocity	Measurement	Velocity
1	850	11	850	21	960	31	800
2	1000	12	810	22	830	32	830
3	740	13	950	23	940	33	850
4	980	14	1000	24	790	34	800
5	900	15	980	25	960	35	880
6	930	16	1000	26	810	36	790
7	1070	17	980	27	940	37	900
8	650	18	960	28	880	38	760
9	930	19	880	29	880	39	840
10	760	20	960	30	880	40	800

A Moving Range chart of the first 20 observations has CL = 122.6; $\hat{\sigma} = \overline{MR2}/d_2 = 122.6/1.128 = 108.7$

$\mu_0 = 734.5; k = 0.5; h = 5$
$K = k\hat{\sigma} = 0.5(108.7) = 54.35$
$H = h\hat{\sigma} = 5(108.7) = 543.5$

MTB>Stat>Control Charts>Time-Weighted Charts>CUSUM

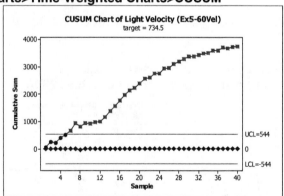

cusum chart: UCL = +544, CL = 0, LCL = -544

The Individuals I-MR chart, with a centerline at $\bar{x} = 909$, displayed a distinct downward trend in measurements, starting at about sample 18. The CUSUM chart reflects a consistent run above the target value 734.5, from virtually the first sample. There is a distinct signal on both charts, of either a trend/drift or a shit in measurements. The out-of-control signals should lead us to investigate and determine the assignable cause.

8-17.   Rework Exercise 8-1 using an EWMA control chart with $\lambda = 0.1$ and $L = 2.7$. Compare your results to those obtained with the cusum.

$\lambda = 0.1$, $L = 2.7$, $\sigma = 25$, CL = μ_0 = 1050, UCL = 1065.49, LCL = 1034.51

MTB > Stat > Control Charts > Time-Weighted Charts > EWMA

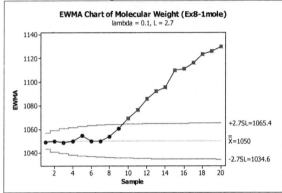

EWMA chart: UCL = 1065.4, CL = 1050, LCL = 1034.6

Process exceeds upper control limit at sample 10; the same as the CUSUM chart.

8-19.   Reconsider the data in Exercise 8-4. Set up an EWMA control chart with $\lambda = 0.2$ and $L = 3$ for this process. Interpret the results.

$\lambda = 0.2$, $L = 3$. From Exercise 8-4, assume $\sigma = 0.05$. CL = μ_0 = 8.02, UCL = 8.07, LCL = 7.97

MTB>Stat>Control Charts>Time-Weighted Charts>EWMA

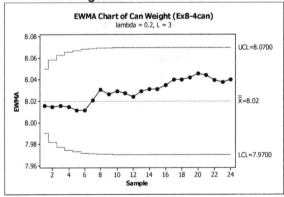

EWMA chart: UCL = 8.0700, CL = 8.02, LCL = 7.9700

The process is in control.

8-21.   Reconsider the data in Exercise 8-7. Apply an EWMA control chart to these data using $\lambda = 0.1$ and $L = 2.7$.

$\lambda = 0.1$, $L = 2.7$, $\hat{\sigma} = 12.16$, CL = $\mu_0 = 950$, UCL = 957.53, LCL = 942.47.

MTB > Stat > Control Charts > Time-Weighted Charts > EWMA

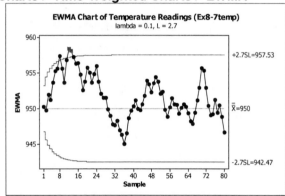

Test Results for EWMA Chart of Ex8-7temp
```
TEST. One point beyond control limits.
Test Failed at points:  12, 13
```

EWMA chart: UCL = 957.53, CL = 950, LCL = 942.47

Process is out of control at samples 8 (beyond upper limit, but not flagged on chart), 12 and 13.

8-23.   Reconsider the data in Exercise 8-8. Set up and apply an EWMA control chart to these data using $\lambda = 0.05$ and $L = 2.6$.

$\lambda = 0.05$, $L = 2.6$, $\hat{\sigma} = 5.634$, CL = $\mu_0 = 175$, UCL = 177.30, LCL = 172.70.

MTB>Stat>Control Charts>Time-Weighted Charts>EWMA

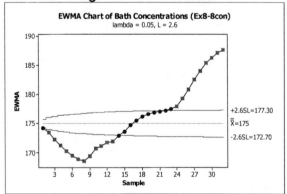

EWMA chart: USL = 177.30, CL = 175, LCL = 172.70

Process is out of control. The process average of $\hat{\mu} = 183.594$ is too far from the process target of $\mu_0 = 175$ for the process variability. The data is grouped into three increasing levels.

8-25. Reconsider the data in Exercise 8-9. Set up and apply an EWMA control chart to these data using $\lambda = 0.1$ and $L = 2.7$.

$\mu_0 = 3200$, $\hat{\sigma} = 5.95$ (from Exercise 8-9), $\lambda = 0.1$, $L = 2.7$

MTB>Stat>Control Charts>Time-Weighted Charts>EWMA

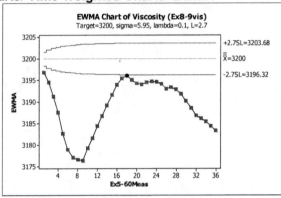

EWMA chart: UCL = 3203.68, CL = 3200, LCL = 3196.32

The process is out of control from the first sample.

8-27. Analyze the data in Exercise 8-4 using a moving average control chart with $w = 5$. Compare the results obtained with the cumulative sum control chart in Exercise 8-4.

$w = 5$, $\mu_0 = 8.02$, $\sigma = 0.05$ (from Exercise 8-4)

MTB>Stat>Control Charts>Time-Weighted Charts>Moving Average

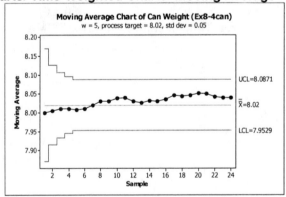

MA chart: UCL = 8.0871, CL = 8.02, LCL = 7.9529

The process is in control, the same result as for Exercise 8-4.

8-29. Show that if the process is in control at the level μ, the exponentially weighted moving average is an unbiased estimator of the process mean.

Assume that t is so large that the starting value $Z_0 = \overline{\overline{x}}$ has no effect.

$$E(Z_t) = E[\lambda \overline{x}_t + (1-\lambda)(Z_{t-1})] = E\left[\lambda \sum_{j=0}^{\infty} (1-\lambda)^j \overline{x}_{t-j}\right] = \lambda \sum_{j=0}^{\infty} (1-\lambda)^j E(\overline{x}_{t-j})$$

Since $E(\overline{x}_{t-j}) = \mu$ and $\lambda \sum_{j=0}^{\infty} (1-\lambda)^j = 1$, $E(Z_t) = \mu$

8-35. An EWMA control chart uses $\lambda = 0.4$. How wide will the limits be on the Shewhart control chart, expressed as a multiple of the width of the steady-state EWMA limits?

$\lambda = 0.4$

For EWMA, steady-state limits are $\pm L\sigma\sqrt{\lambda/(2-\lambda)}$

For Shewhart, steady-state limits are $\pm k\sigma$

$$k\sigma = L\sigma\sqrt{\lambda/(2-\lambda)}$$
$$k = L\sqrt{0.4/(2-0.4)}$$
$$k = 0.5L$$

CHAPTER 9

Other Univariate Statistical Process-Monitoring and Control Techniques

Learning Objectives

After completing this chapter you should be able to:
1. Set up and use $\bar{x}$ and R control charts for short production runs
2. Know how to calculate modified limits for the Shewhart $\bar{x}$ control chart
3. Know how to set up and use an acceptance control chart
4. Use group control charts for multiple-stream processes, and understand the alternative procedures that are available
5. Understand the sources and effects of autocorrelation on standard control charts
6. Know how to sue model-based residuals control charts for autocorrelated data
7. Know how to sue the batch means control chart for autocorrelated data
8. Know what the Cuscore control chart can be used for
9. Know how change point methods relate to statistical process monitoring techniques
10. Understand the practical reason behind the use of adaptive control charts
11. Understand the basis of economic principles of control chart design

Important Terms and Concepts

Acceptance control charts
Autocorrelated process data
Batch means control charts
Control charts on residuals
Deviation from nominal control charts
First-order integrated moving average model
First-order moving average model
Impact of autocorrelation on control charts
Multiple-stream processes
Sample autocorrelation function
Shewhart process model
Time series model

Adaptive control charts
Average run length
Changepoint model for process monitoring
Cuscore statistics and control charts
First-order autoregressive model
First-order mixed model
Group control charts
Modified control charts
Positive autocorrelation
Second-order autoregressive model
Standardized $\bar{x}$ and R control charts

Exercises

Note: Many of the exercises in this chapter were solved using Microsoft Excel 2002, not MINITAB.

9-1.   Use the following data to set up short run $\bar{x}$ and R charts using the DNOM approach. The nominal dimensions for each part are $T_A = 100$, $T_B = 60$, $T_C = 75$, and $T_D = 50$.

Sample Number	Part Type	M_1	M_2	M_3
1	A	105	102	103
2	A	101	98	100
3	A	103	100	99
4	A	101	104	97
5	A	106	102	100
6	B	57	60	59
7	B	61	64	63
8	B	60	58	62
9	C	73	75	77
10	C	78	75	76
11	C	77	75	74
12	C	75	72	79
13	C	74	75	77
14	C	73	76	75
15	D	50	51	49
16	D	46	50	50
17	D	51	46	50
18	D	49	50	53
19	D	50	52	51
20	D	53	51	50

$\hat{\sigma}_A = 2.530$, $n_A = 15$, $\hat{\mu}_A = 101.40$; $\hat{\sigma}_B = 2.297$, $n_B = 9$, $\hat{\mu}_B = 60.444$

$\hat{\sigma}_C = 1.815$, $n_C = 18$, $\hat{\mu}_C = 75.333$; $\hat{\sigma}_D = 1.875$, $n_D = 18$, $\hat{\mu}_D = 50.111$

Standard deviations are approximately the same, so the DNOM chart can be used.

$\bar{R} = 3.8$, $\hat{\sigma} = 2.245$, $n = 3$

$\bar{x}$ chart: CL $= 0.55$, UCL $= 4.44$, LCL $= -3.34$

R chart: CL $= 3.8$, UCL $= D_4\bar{R} = 2.574\,(3.8) = 9.78$, LCL $= 0$

Stat>Control Charts>Variables Charts for Subgroups>Xbar-R Chart

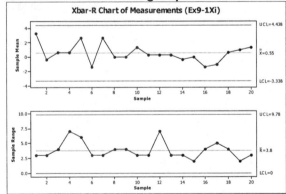

Process is in control, with no samples beyond the control limits or unusual plot patterns.

9-5. A machine has four heads. Samples of $n = 3$ units are selected from each head, and the $\bar{x}$ and R values for an important quality characteristic are computed. The data are shown here. Set up group control charts for this process.

Sample Number	Head 1 $\bar{x}$	Head 1 R	Head 2 $\bar{x}$	Head 2 R	Head 3 $\bar{x}$	Head 3 R	Head 4 $\bar{x}$	Head 4 R
1	53	2	54	1	56	2	55	3
2	51	1	55	2	54	4	54	4
3	54	2	52	5	53	3	57	2
4	55	3	54	3	52	1	51	5
5	54	1	50	2	51	2	53	1
6	53	2	51	1	54	2	52	2
7	51	1	53	2	58	5	54	1
8	52	2	54	4	51	2	55	2
9	50	2	52	3	52	1	51	3
10	51	1	55	1	53	3	53	5
11	52	3	57	2	52	4	55	1
12	51	2	55	1	54	2	58	2
13	54	4	58	2	51	1	53	1
14	53	1	54	4	50	3	54	2
15	55	2	52	3	54	2	52	6
16	54	4	51	1	53	2	58	5
17	53	3	50	2	57	1	53	1
18	52	1	49	1	52	1	49	2
19	51	2	53	3	51	2	50	3
20	52	4	52	2	50	3	52	2

This exercise can be solved in Excel.

Parameter Formulas:

	A	B	C
1	Grand Avg =	=AVERAGE(B14:B33,D14:D33,F14:F33,H14:H33)	
2	Avg R =	=AVERAGE(C14:C33,E14:E33,G14:G33,I14:I33)	
3	s =	4	heads
4	n =	3	units
5	A2 =	1.023	
6	D3 =	0	
7	D4 =	2.574	
8	Xbar UCL =	=B1+B5*B2	
9	Xbar LCL =	=B1-B5*B2	
10	R UCL =	=B7*B2	
11	R LCL =	=B6*B2	

Parameter Values:

	A	B	C
1	Grand Avg =	52.988	
2	Avg R =	2.338	
3	s =	4	heads
4	n =	3	units
5	A2 =	1.023	
6	D3 =	0	
7	D4 =	2.574	
8	Xbar UCL =	55.379	
9	Xbar LCL =	50.596	
10	R UCL =	6.017	
11	R LCL =	0.000	

9-5 continued

Data:

	A	B	C	D	E	F	G	H	I
13	Ex9-5Samp	Ex9-5H1Xb	Ex9-5H1R	Ex9-5H2Xb	Ex9-5H2R	Ex9-5H3Xb	Ex9-5H3R	Ex9-5H4Xb	Ex9-5H4R
14	1	53	2	54	1	56	2	55	3
15	2	51	1	55	2	54	4	54	4
16	3	54	2	52	5	53	3	57	2
17	4	55	3	54	3	52	1	51	5
18	5	54	1	50	3	51	3	53	1

Plot Point Formulas:

	J	K	L	M	N	O	P
13	Ex9-5Xbmax	Ex9-5Xbmin	Ex9-5XbUCL	Ex9-5XbLCL	Ex9-5Rmax	Ex9-5RUCL	Ex9-5RLCL
14	=MAX(B14,D14,F14,H14)	=MIN(B14,D14,F14,H14)	=$B$8	=$B$9	=MAX(C14,E14,G14,I14)	=$B$10	=$B$11
15	=MAX(B15,D15,F15,H15)	=MIN(B15,D15,F15,H15)	=$B$8	=$B$9	=MAX(C15,E15,G15,I15)	=$B$10	=$B$11
16	=MAX(B16,D16,F16,H16)	=MIN(B16,D16,F16,H16)	=$B$8	=$B$9	=MAX(C16,E16,G16,I16)	=$B$10	=$B$11
17	=MAX(B17,D17,F17,H17)	=MIN(B17,D17,F17,H17)	=$B$8	=$B$9	=MAX(C17,E17,G17,I17)	=$B$10	=$B$11
18	=MAX(B18,D18,F18,H18)	=MIN(B18,D18,F18,H18)	=$B$8	=$B$9	=MAX(C18,E18,G18,I18)	=$B$10	=$B$11

Plot Point Values:

	J	K	L	M	N	O	P
13	Ex9-5Xbmax	Ex9-5Xbmin	Ex9-5XbUCL	Ex9-5XbLCL	Ex9-5Rmax	Ex9-5RUCL	Ex9-5RLCL
14	56	53	55.38	50.60	3	6.02	0.00
15	55	51	55.38	50.60	4	6.02	0.00
16	57	52	55.38	50.60	5	6.02	0.00
17	55	51	55.38	50.60	5	6.02	0.00
18	54	50	55.38	50.60	3	6.02	0.00

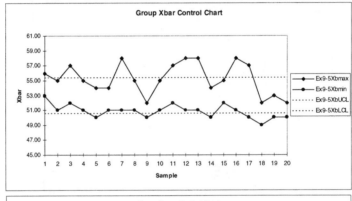

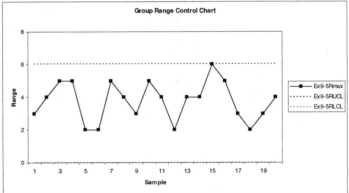

There is no situation where one single head gives the maximum or minimum value of $\overline{x}$ six times in a row. There are many values of $\overline{x}$ max and $\overline{x}$ min that are outside the control limits, so the process is out-of-control. The assignable cause affects all heads, not just a specific one.

9-7.   Reconsider the data in Exercises 9-5 and 9-6. Suppose the process measurements are individual data values, not subgroup averages.

(a) Use observations 1–20 in Exercise 9-5 to construct appropriate group control charts.

     This exercise can be solved in Excel.

Grand Avg =	52.988	
Avg MR =	2.158	
s =	4	heads
n =	2	units
d2 =	1.128	
D3 =	0	
D4 =	3.267	
Xbar UCL =	58.727	
Xbar LCL =	47.248	
R UCL =	7.050	
R LCL =	0.000	

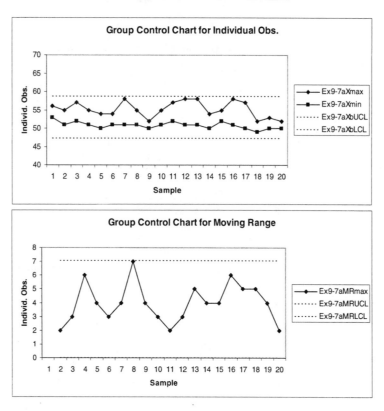

See the discussion in Exercise 9-5.

9-7 continued

(b) Plot observations 21–30 from Exercise 9-6 on the charts from part (a). Discuss your findings.

This exercise can be solved in Excel.

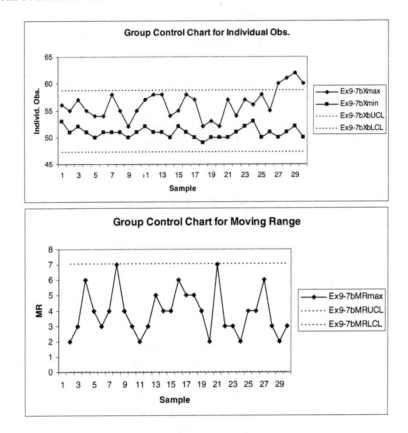

The last four samples from Head 4 remain the maximum of all heads; indicating a potential process change.

9-7 continued

(c) Using observations 1–20, construct an individuals chart using the average of the readings on all four heads as an individual measurement and an *s* control chart using the individual measurements on each head. Discuss how these charts function relative to the group control chart.

Stat>Control Charts>Variables Charts for Subgroups>Xbar-S Chart
Note: Use "Sbar" as the method for estimating standard deviation.

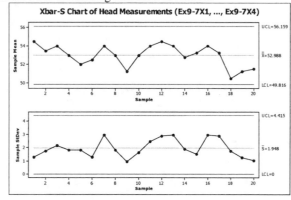

$\bar{x}$ chart: UCL = 56.159, CL = 52.988, LCL = 49.816
s chart: UCL = 4.415, CL = 1.948, LCL = 0

Failure to recognize the multiple stream nature of the process had led to control charts that fail to identify the out-of-control conditions in this process.

(d) Plot observations 21–30 on the control charts from part (c). Discuss your findings.

Stat>Control Charts>Variables Charts for Subgroups>Xbar-S Chart
Note: Use "Sbar" as the method for estimating standard deviation.

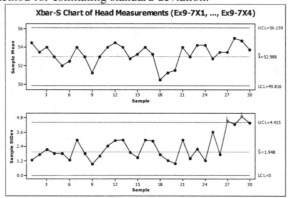

Test Results for S Chart of Ex9-7X1, ..., Ex9-7X4
```
TEST 1. One point more than 3.00 standard deviations from center line.
Test Failed at points:  27, 29
```

Only the *S* chart gives any indication of out-of-control process.

9-9. A sample of five units is taken from a process every half hour. It is known that the process standard deviation is in control with $\sigma = 2.0$. The $\bar{x}$ values for the last 20 samples are below. Specifications on the product are 40 ± 8.

Sample Number	$\bar{x}$	Sample Number	$\bar{x}$
1	41.5	11	40.6
2	42.7	12	39.4
3	40.5	13	38.6
4	39.8	14	42.5
5	41.6	15	41.8
6	44.7	16	40.7
7	39.6	17	42.8
8	40.2	18	43.4
9	41.4	19	42.0
10	43.9	20	41.9

(a) Set up a modified control chart on this process. Use three-sigma limits on the chart and assume that the largest fraction nonconforming that is tolerable is 0.1%.

3-sigma limits

$n = 5$, $\delta = 0.001$, $Z_\delta = Z_{0.001} = 3.090$

$USL = 40 + 8 = 48$, $LSL = 40 - 8 = 32$

$UCL = USL - \left(Z_\delta - 3/\sqrt{n}\right)\sigma = 48 - \left(3.090 - 3/\sqrt{5}\right)(2.0) = 44.503$

$LCL = LSL + \left(Z_\delta - 3/\sqrt{n}\right)\sigma = 32 + \left(3.090 - 3/\sqrt{5}\right)(2.0) = 35.497$

Graph>Time Series Plot>Simple
Note: Reference lines have been used set to the control limit values.

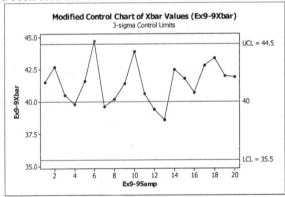

Process is out of control at sample #6.

9-9 continued

(b) Reconstruct the chart in part (a) using two-sigma limits. Is there any difference in the analysis of the data?

   2-sigma limits

$$\text{UCL} = \text{USL} - \left(Z_\delta - 2/\sqrt{n}\right)\sigma = 48 - \left(3.090 - 2/\sqrt{5}\right)(2.0) = 43.609$$

$$\text{LCL} = \text{LSL} + \left(Z_\delta - 2/\sqrt{n}\right)\sigma = 32 + \left(3.090 - 2/\sqrt{5}\right)(2.0) = 36.391$$

Graph > Time Series Plot > Simple
Note: Reference lines have been used set to the control limit values.

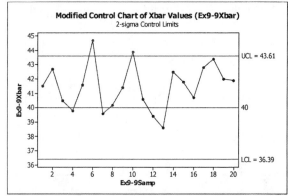

With 3-sigma limits, sample #6 exceeds the UCL, while with 2-sigma limits both samples #6 and #10 exceed the UCL.

(c) Suppose that if the true process fraction nonconforming is 5%, we would like to detect this condition with probability 0.95. Construct the corresponding acceptance control chart.

$$\gamma = 0.05,\ Z_\gamma = Z_{0.05} = 1.645;\quad 1 - \beta = 0.95,\ Z_\beta = Z_{0.05} = 1.645$$

$$\text{UCL} = \text{USL} - \left(Z_\gamma + z_\beta/\sqrt{n}\right)\sigma = 48 - \left(1.645 + 1.645/\sqrt{5}\right)(2.0) = 43.239$$

$$\text{LCL} = \text{LSL} + \left(Z_\gamma + z_\beta/\sqrt{n}\right)\sigma = 32 + \left(1.645 + 1.645/\sqrt{5}\right)(2.0) = 36.761$$

Graph>Time Series Plot>Simple
Note: Reference lines have been used set to the control limit values.

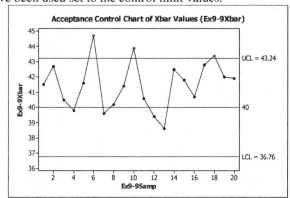

Sample #18 also signals an out-of-control condition.

9-11. Consider a modified control chart with center line at $\mu = 0$, and $\sigma = 1.0$ (known). If $n = 5$, the tolerable fraction nonconforming is $\delta = 0.00135$, and the control limits are at three-sigma, sketch the OC curve for the chart. On the same set of axes, sketch the OC curve corresponding to the chart with two-sigma limits.

$\mu = 0$, $\sigma = 1.0$, $n = 5$, $\delta = 0.00135$, $Z_\delta = Z_{0.00135} = 3.00$

<u>For 3-sigma limits</u>
$Z_\alpha = 3$

$$\text{UCL} = \text{USL} - \left(z_\delta - z_\alpha/\sqrt{n}\right)\sigma = \text{USL} - \left(3.000 - 3/\sqrt{5}\right)(1.0) = \text{USL} - 1.658$$

$$\Pr\{\text{Accept}\} = \Pr\{\bar{x} < \text{UCL}\} = \Phi\left(\frac{\text{UCL} - \mu_0}{\sigma/\sqrt{n}}\right) = \Phi\left(\frac{\text{USL} - 1.658 - \mu_0}{1.0/\sqrt{5}}\right) = \Phi\left((\Delta - 1.658)\sqrt{5}\right); \quad \text{where } \Delta = \text{USL} - \mu_0$$

<u>For 2-sigma limits</u>
$Z_\alpha = 2 \implies \Pr\{\text{Accept}\} = \Phi\left((\Delta - 2.106)\sqrt{5}\right)$

$$p = \Pr\{x > \text{USL}\} = 1 - \Pr\{x \le \text{USL}\} = 1 - \Phi\left(\frac{\text{USL} - \mu_0}{\sigma}\right) = 1 - \Phi(\Delta)$$

This exercise can be solved in Excel.

	A	B	C	D	E
1	DELTA=USL-mu0	CumNorm(DELTA)	p	Pr(Accept@3)	Pr(Accept@2)
2	3.5	=NORMDIST(A2,0,1,TRUE)	=1-B2	=NORMDIST(A2,1.658,(1/SQRT(5)),TRUE)	=NORMDIST(A2,2.106,1/SQRT(5),TRUE)
3	3.25	=NORMDIST(A3,0,1,TRUE)	=1-B3	=NORMDIST(A3,1.658,(1/SQRT(5)),TRUE)	=NORMDIST(A3,2.106,1/SQRT(5),TRUE)
4	3	=NORMDIST(A4,0,1,TRUE)	=1-B4	=NORMDIST(A4,1.658,(1/SQRT(5)),TRUE)	=NORMDIST(A4,2.106,1/SQRT(5),TRUE)
5	2.5	=NORMDIST(A5,0,1,TRUE)	=1-B5	=NORMDIST(A5,1.658,(1/SQRT(5)),TRUE)	=NORMDIST(A5,2.106,1/SQRT(5),TRUE)
6	...				

	A	B	C	D	E
1	DELTA=USL-mu0	CumNorm(DELTA)	p	Pr(Accept@3)	Pr(Accept@2)
2	3.50	0.9998	0.0002	1.0000	0.9991
3	3.25	0.9994	0.0006	0.9998	0.9947
4	3.00	0.9987	0.0013	0.9987	0.9772
5	2.50	0.9938	0.0062	0.9701	0.8108
6	...				

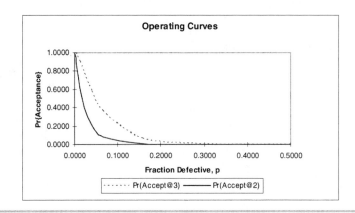

9-13.   An $\bar{x}$ chart is to be designed for a quality characteristic assumed to be normal with a standard deviation of 4. Specifications on the product quality characteristics are 50 ± 20. The control chart is to be designed so that if the fraction nonconforming is 1%, the probability of a point falling inside the control limits will be 0.995. The sample size is $n = 4$. What are the control limits and center line for the chart?

Design a modified control chart.

$n = 4$, USL $= 70$, LSL $= 30$, $S = 4$

$\delta = 0.01$, $Z_\delta = 2.326$

$1 - \alpha = 0.995$, $\alpha = 0.005$, $Z_\alpha = 2.576$

$$\text{UCL} = \text{USL} - \left(Z_\delta - Z_\alpha/\sqrt{n}\right)\sigma = (50+20) - \left(2.326 - 2.576/\sqrt{4}\right)(4) = 65.848$$

$$\text{LCL} = \text{LSL} + \left(Z_\delta - Z_\alpha/\sqrt{n}\right)\sigma = (50-20) + \left(2.326 - 2.576/\sqrt{4}\right)(4) = 34.152$$

9-15.   A normally distributed quality characteristic is controlled by $\bar{x}$ and R charts having the following parameters ($n = 4$, both charts are in control):

R Chart	$\bar{x}$ Chart
UCL $= 18.795$	UCL $= 626.00$
Center line $= 8.236$	Center line $= 620.00$
LCL $= 0$	LCL $= 614.00$

(a) What is the estimated standard deviation of the quality characteristic x?

$n = 4$, $\bar{R} = 8.236$, $\bar{\bar{x}} = 620.00$

$\hat{\sigma}_x = \bar{R}/d_2 = 8.236/2.059 = 4.000$

(b) If specifications are 610 ± 15, what is your estimate of the fraction of nonconforming material produced by this process when it is in control at the given level?

$\hat{p} = \Pr\{x < \text{LSL}\} + \Pr\{x > \text{USL}\}$

$\quad = \Pr\{x < 595\} + \left[1 - \Pr\{x \le 625\}\right]$

$\quad = \Phi\left(\dfrac{595-620}{4.000}\right) + \left[1 - \Phi\left(\dfrac{625-620}{4.000}\right)\right]$

$\quad = 0.0000 + \left[1 - 0.8944\right] = 0.1056$

(c) Suppose you wish to establish a modified $\bar{x}$ chart to substitute for the original $\bar{x}$ chart. The process mean is to be controlled so that the fraction nonconforming is less than 0.005. The probability of type I error is to be 0.01. What control limits do you recommend?

$\delta = 0.005$, $Z_\delta = Z_{0.005} = 2.576$

$\alpha = 0.01$, $Z_\alpha = Z_{0.01} = 2.326$

$$\text{UCL} = \text{USL} - \left(Z_\delta - Z_\alpha/\sqrt{n}\right)\sigma = 625 - \left(2.576 - 2.326/\sqrt{4}\right)4 = 619.35$$

$$\text{LCL} = \text{LSL} + \left(Z_\delta - Z_\alpha/\sqrt{n}\right)\sigma = 595 + \left(2.576 - 2.326/\sqrt{4}\right)4 = 600.65$$

9-17.   Consider the molecular weight data in Exercise 9-16. Construct a cusum control chart on the residuals from the model you fit to the data in part (c) of that exercise.

Let $\mu_0 = 0$, $\delta = 1$ sigma, $k = 0.5$, $h = 5$.

Stat>Control Charts>Time-Weighted Charts>CUSUM

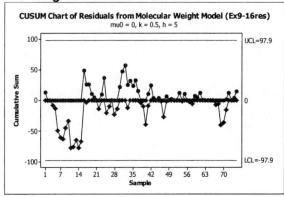

cusum chart: UCL = 97.9, CL = 0, LCL = -97.9

No observations exceed the control limit. The residuals are in control.

9-19.   Set up a moving center-line EWMA control chart for the molecular weight data in Exercise 9-16. Compare it to the residual control chart in Exercise 9-16, part (c).

2048	2039	2051	2002	2029
2025	2015	2056	1967	2019
2017	2021	2018	1994	2016
1995	2010	2030	2001	2010
1983	2012	2023	2013	2006
1943	2003	2036	2016	2009
1940	1979	2019	2019	1990
1947	2006	2000	2036	1986
1972	2042	1986	2015	1947
1983	2000	1952	2032	1958
1935	2002	1988	2016	1983
1948	2010	2016	2000	2010
1966	1975	2002	1988	2000
1954	1983	2004	2010	2015
1970	2021	2018	2015	2032

To find the optimal λ, fit an ARIMA $(0,1,1)$ (= EWMA = IMA$(1,1)$): **Stat > Time Series > ARIMA**

```
ARIMA Model: Ex9-16mole
Final Estimates of Parameters
Type        Coef   SE Coef      T       P
MA   1    0.0762    0.1181    0.65   0.521 ****
Constant  -0.211    2.393    -0.09   0.930
...
```

$\lambda = 1 - MA1 = 1 - 0.0762 = 0.9238$

To find an estimate of standard deviation, construct a Moving Range chart and use $CL = \overline{MR}$

$\hat{\sigma} = \overline{MR}/d_2 = 17.97/1.128 = 15.93$

9-19 continued

A moving centerline EWMA chart can be constructed in Excel.

lambda =	0.9238	sigma^ =		15.934			
t	xt	zt	CL	UCL	LCL	OOC?	
0		=AVERAGE(B5:B80)					
1	2048	=$B$1*B5+(1-$B$1)*C4	=C4	=D5+3*$D$1	=D5-3*$D$1	=IF(B5>E5,"above UCL",IF(B5<F5,"below LCL","No"))	
=A5+1	2025	=$B$1*B6+(1-$B$1)*C5	=C5	=D6+3*$D$1	=D6-3*$D$1	=IF(B6>E6,"above UCL",IF(B6<F6,"below LCL","No"))	
=A6+1	2017	=$B$1*B7+(1-$B$1)*C6	=C6	=D7+3*$D$1	=D7-3*$D$1	=IF(B7>E7,"above UCL",IF(B7<F7,"below LCL","No"))	
=A7+1	1995	=$B$1*B8+(1-$B$1)*C7	=C7	=D8+3*$D$1	=D8-3*$D$1	=IF(B8>E8,"above UCL",IF(B8<F8,"below LCL","No"))	
=A8+1	1983	=$B$1*B9+(1-$B$1)*C8	=C8	=D9+3*$D$1	=D9-3*$D$1	=IF(B9>E9,"above UCL",IF(B9<F9,"below LCL","No"))	
...							

lambda =	0.9238	sigma^ =	15.93			
t	xt	zt	CL	UCL	LCL	OOC?
0		2000.947				
1	2048	2044.415	2000.947	2048.749	1953.145	No
2	2025	2026.479	2044.415	2092.217	1996.613	No
3	2017	2017.722	2026.479	2074.281	1978.677	No
4	1995	1996.731	2017.722	2065.524	1969.920	No
5	1983	1984.046	1996.731	2044.533	1948.929	No
...						
6	1943	1946.128	1984.046	2031.848	1936.244	No
7	1940	1940.467	1946.128	1993.930	1898.326	No
8	1947	1946.502	1940.467	1988.269	1892.665	No
9	1972	1970.057	1946.502	1994.304	1898.700	No
10	1983	1982.014	1970.057	2017.859	1922.255	No
11	1935	1938.582	1982.014	2029.816	1934.212	No
12	1948	1947.282	1938.582	1986.384	1890.780	No
13	1966	1964.574	1947.282	1995.084	1899.480	No
14	1954	1954.806	1964.574	2012.376	1916.772	No
15	1970	1968.842	1954.806	2002.608	1907.004	No
16	2039	2033.654	1968.842	2016.644	1921.040	above UCL

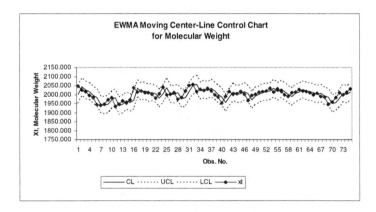

Observation 6 exceeds the upper control limit compared to one out-of-control signal at observation 16 on the Individuals control chart.

9-21.   Consider the concentration data in Exercise 9-20. Construct a cusum chart in the residuals from the model you fit in part (c) of that exercise.

Let $\mu_0 = 0$, $\delta = 1$ sigma, $k = 0.5$, $h = 5$.

Stat>Control Charts>Time-Weighted Charts>CUSUM

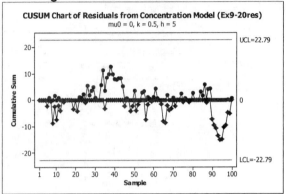

cusum chart: UCL = +22.79, CL = 0, LCL = -22.79

No observations exceed the control limit. The residuals are in control, and the AR(1) model for concentration should be a good fit.

9-23.   Set up a moving center-line EWMA control chart for the concentration data in Exercise 9-20. Compare it to the residuals control chart in Exercise 9-20, part (c).

204	190	208	207	200
202	196	209	204	202
201	199	209	201	202
202	203	206	197	207
197	199	200	189	206
201	207	203	189	211
198	204	202	196	205
188	207	195	193	210
195	209	196	193	210
189	205	203	198	198
195	202	196	194	194
192	200	197	198	192
196	208	197	199	189
194	214	203	204	188
196	205	205	200	189
199	211	194	203	194
197	212	199	200	194
197	214	201	197	198
192	210	198	196	196
195	208	202	202	200

To find the optimal λ, fit an ARIMA $(0,1,1)$ (= EWMA = IMA(1,1)): **Stat > Time Series > ARIMA**

```
ARIMA Model: Ex9-20conc
Final Estimates of Parameters
Type          Coef   SE Coef      T       P
MA     1    0.2945   0.0975    3.02   0.003 ****
Constant  -0.0452   0.3034   -0.15   0.882
...
```

$\lambda = 1 - MA1 = 1 - 0.2945 = 0.7055$

9-21 continued

To find an estimate of standard deviation, construct a Moving Range chart and use $CL = \overline{MR}$

$$\hat{\sigma} = \overline{MR}/d_2 = 3.64/1.128 = 3.227$$

A moving centerline EWMA chart may be constructed in Excel.

lamda =	0.7055	sigma^ =		3.227			
t	xt	zt	CL	UCL =	LCL =	OOC?	
0		=AVERAGE(B5:B105)					
1	204	=$B$1*B5+(1-$B$1)*C4	=C4	=D5+3*$D$1	=D5-3*$D$1	=IF(B5>E5,"above UCL",IF(B5<F5,"below LCL",0))	
=A5+1	202	=$B$1*B6+(1-$B$1)*C5	=C5	=D6+3*$D$1	=D6-3*$D$1	=IF(B6>E6,"above UCL",IF(B6<F6,"below LCL",0))	
=A6+1	201	=$B$1*B7+(1-$B$1)*C6	=C6	=D7+3*$D$1	=D7-3*$D$1	=IF(B7>E7,"above UCL",IF(B7<F7,"below LCL",0))	
=A7+1	202	=$B$1*B8+(1-$B$1)*C7	=C7	=D8+3*$D$1	=D8-3*$D$1	=IF(B8>E8,"above UCL",IF(B8<F8,"below LCL",0))	
=A8+1	197	=$B$1*B9+(1-$B$1)*C8	=C8	=D9+3*$D$1	=D9-3*$D$1	=IF(B9>E9,"above UCL",IF(B9<F9,"below LCL",0))	
...							

lamda =	0.706	sigma^ =	3.23			
t	xt	zt	CL	UCL =	LCL =	OOC?
0		200.010				
1	204	202.825	200.010	209.691	190.329	0
2	202	202.243	202.825	212.506	193.144	0
3	201	201.366	202.243	211.924	192.562	0
4	202	201.813	201.366	211.047	191.685	0
5	197	198.418	201.813	211.494	192.132	0
6	201	200.239	198.418	208.099	188.737	0
7	198	198.660	200.239	209.920	190.558	0
8	188	191.139	198.660	208.341	188.979	below LCL
...						

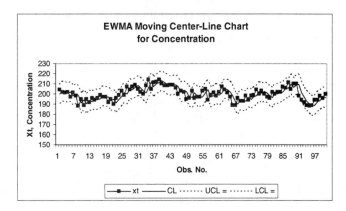

EWMA Moving Center-Line Chart for Concentration

The control chart of concentration data signals out of control at three observations (8, 56, 90).

9-25. Consider the temperature data in Exercise 9-24. Set up a cusum control chart on the residuals from the model you fit to the data in part (c) of that exercise. Compare it to the individuals chart you constructed using the residuals. (In Exercise 9-24 part (c), a first-order autoregressive model was fit to the temperature data.)

MTB>Stat>Control Charts>Time-Weighted Charts>CUSUM

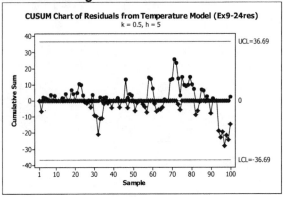

cusum chart: UCL = +36.69, CL = 0, LCL = -36.69

No observations exceed the control limits. The residuals are in control, indicating the process is in control. This is the same conclusion as applying an Individuals control chart to the model residuals.

9-27. Set up a moving center-line EWMA control chart for the temperature data in Exercise 9-24. Compare it to the residuals control chart in Exercise 9-24, part (c).

491	526	489	502	528
482	533	496	494	513
490	533	489	492	511
495	527	494	490	512
499	520	496	489	522
499	514	514	495	523
507	517	505	498	517
503	508	511	501	522
510	515	513	518	518
509	501	508	521	505
510	497	498	535	510
510	483	500	533	508
515	491	502	524	510
513	489	506	515	487
520	496	500	529	481
518	501	495	525	483
517	496	489	526	473
526	495	509	528	479
525	488	511	534	475
519	491	508	530	484

To find the optimal λ, fit an ARIMA $(0,1,1)$ (= EWMA = IMA$(1,1)$): **Stat>Time Series>ARIMA**

```
ARIMA Model: Ex9-24temp
Final Estimates of Parameters
Type        Coef   SE Coef      T      P
MA    1   0.0794   0.1019    0.78  0.438  ****
Constant -0.0711   0.6784   -0.10  0.917
...
```

$\lambda = 1 - MA1 = 1 - 0.0794 = 0.9206$

9-27 continued

To find an estimate of standard deviation, construct a Moving Range chart and use $CL=\overline{MR}$

$$\hat{\sigma} = \overline{MR}/d_2 = 5.75/1.128 = 5.0975$$

A moving centerline EWMA chart may be constructed in Excel.

		lambda =		0.9206	sigma^ =		=5.75/1.128	
t	xt	zt	CL	UCL	LCL			OOC?
0		=AVERAGE(B5:B105)						
1	491	=$D$1*B5+(1-$D$1)*C4	=C4	=D5+3*$F$1	=D5-3*$F$1			=IF(B5>E5,"above UCL",IF(B5<F5,"below LCL",0))
=A5+1	482	=$D$1*B6+(1-$D$1)*C5	=C5	=D6+3*$F$1	=D6-3*$F$1			=IF(B6>E6,"above UCL",IF(B6<F6,"below LCL",0))
=A6+1	490	=$D$1*B7+(1-$D$1)*C6	=C6	=D7+3*$F$1	=D7-3*$F$1			=IF(B7>E7,"above UCL",IF(B7<F7,"below LCL",0))
=A7+1	495	=$D$1*B8+(1-$D$1)*C7	=C7	=D8+3*$F$1	=D8-3*$F$1			=IF(B8>E8,"above UCL",IF(B8<F8,"below LCL",0))
=A8+1	499	=$D$1*B9+(1-$D$1)*C8	=C8	=D9+3*$F$1	=D9-3*$F$1			=IF(B9>E9,"above UCL",IF(B9<F9,"below LCL",0))
...								

		lambda =	0.921	sigma^ =	5.098	
t	xt	zt	CL	UCL	LCL	OOC?
0		506.520				
1	491	492.232	506.520	521.813	491.227	below LCL
2	482	482.812	492.232	507.525	476.940	0
3	490	489.429	482.812	498.105	467.520	0
4	495	494.558	489.429	504.722	474.137	0
5	499	498.647	494.558	509.850	479.265	0
...						

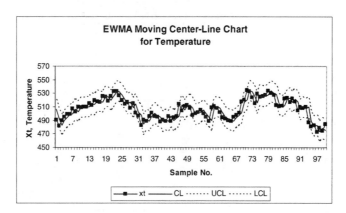

A few observations exceed the upper limit (46, 58, 69) and the lower limit (1, 94), similar to the two out-of-control signals on the Individuals control chart (71, 94).

9-29. The viscosity of a chemical product is read every 2 minutes. Some data from this process are shown in the table on the next page (read down, then across from left to right).

29.330	33.220	27.990	24.280
19.980	30.150	24.130	22.690
25.760	27.080	29.200	26.600
29.000	33.660	34.300	28.860
31.030	36.580	26.410	28.270
32.680	29.040	28.780	28.170
33.560	28.080	21.280	28.580
27.500	30.280	21.710	30.760
26.750	29.350	21.470	30.620
30.550	33.600	24.710	20.840
28.940	30.290	33.610	16.560
28.500	20.110	36.540	25.230
28.190	17.510	35.700	31.790
26.130	23.710	33.680	32.520
27.790	24.220	29.290	30.280
27.630	32.430	25.120	26.140
29.890	32.440	27.230	19.030
28.180	29.390	30.610	24.340
26.650	23.450	29.060	31.530
30.010	23.620	28.480	31.950
30.800	28.120	32.010	31.680
30.450	29.940	31.890	29.100
36.610	30.560	31.720	23.150
31.400	32.300	29.090	26.740
30.830	31.580	31.920	32.440

(a) Is there a serious problem with autocorrelation in these data?
Stat>Time Series>Autocorrelation

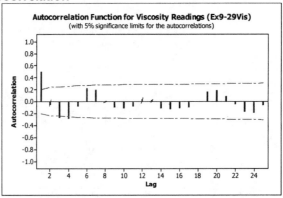

Autocorrelation Function: Ex9-29Vis

Lag	ACF	T	LBQ
1	0.494137	4.94	25.16
2	-0.049610	-0.41	25.41
3	-0.264612	-2.17	32.78
4	-0.283150	-2.22	41.29
5	-0.071963	-0.54	41.85
...			

$r_1 = 0.49$, indicating a strong positive correlation at lag 1. There is a serious problem with autocorrelation in viscosity readings.

9-29 continued

(b) Set up a control chart for individuals with a moving range used to estimate process variability. What conclusion can you draw from this chart?

Stat>Control Charts>Variables Charts for Individuals>Individuals

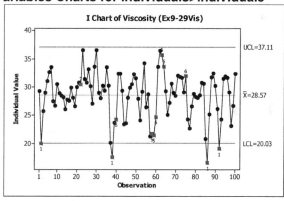

Test Results for I Chart of Ex9-29Vis
```
TEST 1. One point more than 3.00 standard deviations from center line.
Test Failed at points:  2, 38, 86, 92
TEST 5. 2 out of 3 points more than 2 standard deviations from center line (on
        one side of CL).
Test Failed at points:  38, 58, 59, 63, 86
TEST 6. 4 out of 5 points more than 1 standard deviation from center line (on
        one side of CL).
Test Failed at points:  40, 60, 64, 75
TEST 7. 15 points within 1 standard deviation of center line (above and below
        CL).
Test Failed at points:  22
TEST 8. 8 points in a row more than 1 standard deviation from center line
        (above and below CL).
Test Failed at points:  64
```

x chart: UCL = 37.11, CL = 28.57, LCL = 20.03

Process is out of control, violating many of the tests for special causes. The viscosity measurements appear to wander over time.

9-29 continued

(c) Design a cusum control scheme for this process, assuming that the observations are uncorrelated. How does the cusum perform?

Let target = $\mu_0 = 28.569$, $k = 0.5$, $h = 5$

MTB>Stat>Control Charts>Time-Weighted Charts>CUSUM

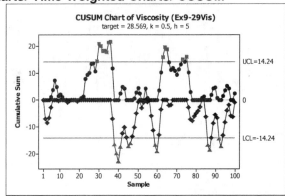

cusum chart: UCL = +14.24, CL = 0, LCL = -14.24

Several observations are out of control on both the lower and upper sides.

(d) Set up an EWMA control chart with $\lambda = 0.15$ for the process. How does this chart perform?

Let $\lambda = 0.15$ and $L = 2.7$.

MTB>Stat>Control Charts>Time-Weighted Charts>CUSUM

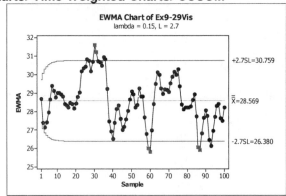

EWMA chart: UCL = 30.759, CL = 28.569, LCL = 26.380

The process is not in control. There are wide swings in the plot points and several are beyond the control limits.

9-29 continued

(e) Set up a moving center-line EWMA scheme for these data.

To find the optimal λ, fit an ARIMA $(0,1,1)$ $(= \text{EWMA} = \text{IMA}(1,1))$: **Stat>Time Series>ARIMA**

```
ARIMA Model: Ex9-29Vis
Final Estimates of Parameters
Type          Coef  SE Coef       T      P
MA    1    -0.1579   0.1007   -1.57  0.120  ****
Constant   0.0231   0.4839    0.05  0.962
...
```

$\lambda = 1 - \text{MA1} = 1 - (-0.1579) = 1.1579$

To find an estimate of standard deviation, construct a Moving Range chart and use $\text{CL} = \overline{\text{MR}}$

$\hat{\sigma} = \overline{\text{MR}} / d_2 = 3.21 / 1.128 = 2.8457$

A moving centerline EWMA chart may be constructed in Excel.

	A	B	C	D	E	F	G
1			lambda =	1.1579	sigma^ =	=3.212/1.128	
2							
3	I	Xi	Zi	CL	UCL	LCL	OOC?
4	0		=AVERAGE(B5:B105)				
5	1	29.33	=$D$1*B5+(1-$D$1)*C4	=C4	=D5+3*$F$1	=D5-3*$F$1	=IF(B5>E5,"above UCL",IF(B5<F5,"below LCL",0))
6	2	19.98	=$D$1*B6+(1-$D$1)*C5	=C5	=D6+3*$F$1	=D6-3*$F$1	=IF(B6>E6,"above UCL",IF(B6<F6,"below LCL",0))
7	3	25.76	=$D$1*B7+(1-$D$1)*C6	=C6	=D7+3*$F$1	=D7-3*$F$1	=IF(B7>E7,"above UCL",IF(B7<F7,"below LCL",0))
8	4	29	=$D$1*B8+(1-$D$1)*C7	=C7	=D8+3*$F$1	=D8-3*$F$1	=IF(B8>E8,"above UCL",IF(B8<F8,"below LCL",0))
9	5	31.03	=$D$1*B9+(1-$D$1)*C8	=C8	=D9+3*$F$1	=D9-3*$F$1	=IF(B9>E9,"above UCL",IF(B9<F9,"below LCL",0))
10	...						

	A	B	C	D	E	F	G
1			lambda =	1.158	sigma^ =	2.85	
2							
3	I	Xi	Zi	CL	UCL	LCL	OOC?
4	0		28.479				
5	1	29.330	29.464	28.479	37.022	19.937	0
6	2	19.980	18.482	29.464	38.007	20.922	below LCL
7	3	25.760	26.909	18.482	27.025	9.940	0
8	4	29.000	29.330	26.909	35.452	18.367	0
9	5	31.030	31.298	29.330	37.873	20.788	0
10	...						

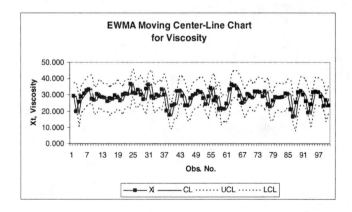

A few observations exceed the upper limit (87) and the lower limit (2, 37, 55, 85).

9-29 continued

(f) Suppose that a reasonable model for the viscosity data is an AR (2) model. How could this model be used to assist in the development of a statistical process control procedure for viscosity? Set up an appropriate control chart and use it to assess the current state of statistical control.

To develop an SPC procedure for viscosity, fit an ARIMA model with two autoregressive parameters to the original data, and then create a control scheme based on residuals from this model. An Individuals control chart would be appropriate for independent, normally-distributed residuals.

Stat>Time Series>ARIMA

ARIMA Model: Ex9-29Vis
```
Final Estimates of Parameters
Type        Coef  SE Coef       T      P
AR    1   0.7193   0.0923    7.79  0.000   ****
AR    2  -0.4349   0.0922   -4.72  0.000   ****
Constant  20.5017   0.3278   62.54  0.000
Mean      28.6514   0.4581
...
```

Stat>Control Charts>Variables Charts for Individuals>Individuals

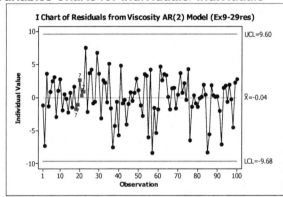

Test Results for I Chart of Ex9-29res
```
TEST 7. 15 points within 1 standard deviation of center line (above and below
     CL).
Test Failed at points:  18, 19, 20, 21, 22
```

x chart: UCL = +9.60, CL = 0.04, LCL = -9.68

The model residuals signal a potential issue with viscosity around observation 20. Otherwise the process appears to be in control, with a good distribution of points between the control limits and no patterns.

9-31.   An $\bar{x}$ chart is used to maintain current control of a process. The cost parameters are $a_1 = \$0.50$, $a_2 = \$0.10$, $a_3 = \$25$, $a'_3 = \$50$, and $a_4 = \$100$. A single assignable cause of magnitude $\delta = 2$ occurs, and the duration of the process in control is an exponential random variable with mean 100 h. Sampling and testing require 0.05 h, and it takes 2 h to locate the assignable cause. Assume that equation 9-31 is the appropriate process model.

(a) Evaluate the cost of the arbitrary control chart design $n = 5$, $k = 3$, and $h = 1$.

$\lambda = 0.01/\text{hr}$ or $1/\lambda = 100\text{hr}$; $\delta = 2.0$

$a_1 = \$0.50/\text{sample}$; $a_2 = \$0.10/\text{unit}$; $a'_3 = \$50$; $a_3 = \$25$; $a_4 = \$100/\text{hr}$

$g = 0.05\text{hr/sample}$; $D = 2\text{hr}$

$n = 5$, $k = 3$, $h = 1$, $\alpha = 0.0027$

$$\beta = \Phi\left(\frac{(\mu_0 + k\sigma/\sqrt{n}) - (\mu_0 + \delta\sigma)}{\sigma/\sqrt{n}}\right) - \Phi\left(\frac{(\mu_0 - k\sigma/\sqrt{n}) - (\mu_0 + \delta\sigma)}{\sigma/\sqrt{n}}\right)$$

$$= \Phi\left(k - \delta\sqrt{n}\right) - \Phi\left(-k - \delta\sqrt{n}\right) = \Phi\left(3 - 2\sqrt{5}\right) - \Phi\left(-3 - 2\sqrt{5}\right)$$

$$= \Phi(-1.472) - \Phi(-7.472) = 0.0705 - 0.0000 = 0.0705$$

$$\tau \cong \frac{h}{2} - \frac{\lambda h^2}{12} = \frac{1}{2} - \frac{0.01(1^2)}{12} = 0.4992$$

$$\frac{\alpha e^{-\lambda h}}{\left(1 - e^{-\lambda h}\right)} \cong \frac{\alpha}{\lambda h} = \frac{0.0027}{0.01(1)} = 0.27$$

$E(L) = \$4.12/\text{hr}$

This answer can be calculated in Excel.

Parameters and Formulas:

	A	B	C	D	E	F
1	lambda	delta	a1, $/sample	a2, $/unit	a3', $	a3, $
2	=1/100	2	0.5	0.1	50	25
3						
4	Part (a)					
5	n	k	h	alpha	beta	1-beta
6	5	3	1	0.0027	=NORMDIST(B6-$B$2*SQRT(A6),0,1,TRUE) -NORMDIST(-B6-$B$2*SQRT(A6),0,1,TRUE)	=1-E6

	G	H	I	J	K	L
1	a4, $/hr	g, hr/sample	D, hr			
2	100	0.05	2			
3						
4						
5	approx tau	approx alpha+	1st	num	denom	E(L), $/hr
6	=C6/2-$A$2*C6^2/12	=D6/($A$2*C6)	=($C$2+$D$2*A6)/C6	=$G$2*(C6/F6-G6+$H$2*A6+$I$2)+$F$2+$E$2*H6	=1/$A$2+C6/F6-G6+$H$2*A6+$I$2	=I6+J6/K6

Results:

	A	B	C	D	E	F	G	H	I	J	K	L
1	lambda	delta	a1, $/sample	a2, $/unit	a3', $	a3, $	a4, $/hr	g, hr/sample	D, hr			
2	0.01	2	$0.50	$0.10	$50	$25	$100	0.05	2			
3												
4	Part (a)											
5	n	k	h	alpha	beta	1-beta	approx tau	approx alpha+	1st	num	denom	E(L), $/hr
6	5	3	1	0.0027	0.070492119	0.9295	0.4992	0.27	1	321.1671441	102.8266714	4.12

9-31 continued

(b) Evaluate the cost of the arbitrary control chart design $n = 5$, $k = 3$, and $h = 0.5$.

$n = 5$, $k = 3$, $h = 0.5$, $\alpha = 0.0027$, $\beta = 0.0705$

$$\tau \cong \frac{h}{2} - \frac{\lambda h^2}{12} = \frac{0.5}{2} - \frac{0.01(0.5^2)}{12} = 0.2498$$

$$\frac{\alpha e^{-\lambda h}}{\left(1 - e^{-\lambda h}\right)} \cong \frac{\alpha}{\lambda h} = \frac{0.0027}{0.01(0.5)} = 0.54$$

$E(L) = \$4.98/\text{hr}$

	A	B	C	D	E	F	G	H	I	J	K	L
1	lambda	delta	a1, $/sample	a2, $/unit	a3', $	a3, $	a4, $/hr	g, hr/sample	D, hr			
2	0.01	2	$0.50	$0.10	$50	$25	$100	0.05	2			
3												
8	Part (b)											
9	n	k	h		alpha	beta	1-beta	approx tau	approx alpha+	1st	num	denom	E(L), $/hr
10	5	3	0.5		0.0027	0.070492119	0.9295	0.2498	0.54	2	305.8127387	102.5381274	4.98

(c) Determine the economically optimum design.

$n = 5$, $k_{opt} = 3.080$, $h_{opt} = 1.368$, $\alpha = 0.00207$, $1 - \beta = 0.918$

$E(L) = \$4.01392/\text{hr}$

9-33.   An $\bar{x}$ chart is used to maintain current control of a process. The cost parameters are $a_1 = \$2$, $a_2 = \$0.50$, $a_3 = \$50$, $a'_3 = \$75$, and $a_4 = \$200$. A single assignable cause occurs, with magnitude $\delta = 1$, and the run length of the process in control is exponentially distributed with mean 100 h. It requires 0.05 h to sample and test, and 1 h to locate the assignable cause.

(a) Evaluate the cost of the arbitrary $\bar{x}$ chart design $n = 5$, $k = 3$, and $h = 0.5$.

$\lambda = 0.01/\text{hr}$ or $1/\lambda = 100\text{hr}$

$\delta = 2.0$

$a_1 = \$2/\text{sample}$

$a_2 = \$0.50/\text{unit}$

$a'_3 = \$75$

$a_3 = \$50$

$a_4 = \$200/\text{hr}$

$g = 0.05 \text{ hr/sample}$

$D = 1 \text{ hr}$

$n = 5$, $k = 3$, $h = 0.5$, $\alpha = 0.0027$

$$\beta = \Phi\left(k - \delta\sqrt{n}\right) - \Phi\left(-k - \delta\sqrt{n}\right) = \Phi\left(3 - 1\sqrt{5}\right) - \Phi\left(-3 - 1\sqrt{5}\right)$$

$$= \Phi(-1.472) - \Phi(-7.472) = 0.775 - 0.0000 = 0.775$$

$$\tau \cong \frac{h}{2} - \frac{\lambda h^2}{12} = \frac{0.5}{2} - \frac{0.01(0.5^2)}{12} = 0.2498$$

$$\frac{\alpha e^{-\lambda h}}{\left(1 - e^{-\lambda h}\right)} \cong \frac{\alpha}{\lambda h} = \frac{0.0027}{0.01(0.5)} = 0.54$$

$E(L) = \$16.17/\text{hr}$

9-33 continued

This answer can be calculated in Excel.

Parameters and Formulas:

lambda	delta	a1,$/sample	a2,$/unit	a3', $		a3, $
=1/100	1	2	0.5	75		50
n	k	h	alpha	beta		1-beta
5	3	0.5	0.0027	=NORMDIST(B6-$B$2*SQRT(A6),0,1,TRUE) -NORMDIST(-B6-$B$2*SQRT(A6),0,1,TRUE)		=1-E6

a4, $/hr	g,hr/sample	D, hr			
200	0.05	1			
approx tau	approx alpha+	1st	num		denom
=C6/2-$A$2*C6^2/12	=D6/($A$2*C6)	=($C$2+$D$2*A6)/C6	=$G$2*(C6/F6-G6+$H$2*A6+$I$2)+$F$2+$E$2*H6		=1/$A$2+C6/F6-G6+$H$2*A6+$I$2

Results:

lambda	delta	a1,$/sample	a2,$/unit	a3', $	a3, $	a4, $/hr	g,hr/sample	D, hr			
0.01	1	$2	$0.50	$75	$50	$200	0.05	1			
n	k	h	alpha	beta	1-beta	approx tau	approx alpha+	1st	num	denom	E(L), $/hr
5	3	0.5	0.0027	0.777546112	0.222453888	0.2498	0.54	9	740.0730326	103.2478652	16.17

(b) Find the economically optimum design.

$n = 10$, $k_{opt} = 2.240$, $h_{opt} = 2.489018$, $\alpha = 0.025091$, $1 - \beta = 0.8218083$

$E(L) = \$10.39762/hr$

CHAPTER 10

Multivariate Process Monitoring and Control

Learning Objectives

After completing this chapter you should be able to:

1. Understand why applying several univariate control carts simultaneously to a set of related quality characteristics may be an unsatisfactory monitoring procedure
2. How the multivariate normal distribution is used as a model for multivariate process data
3. Know how to estimate the mean vector and covariance matrix from a sample of multivariate observations
4. Know how to set up and use a chi-square control chart
5. Know how to set up and use the Hotelling T^2 control chart
6. Know how to set up and use the multivariate exponentially weighted moving average (MEWMA) control chart
7. Know how to use multivariate control charts for individual observations
8. Know how to find the phase I and phase II limits for multivariate control charts
9. Use control charts for monitoring multivariate variability
10. Understand the basis of the regression adjustment procedure and know how to apply regression adjustment in process monitoring
11. Understand the basis of principal components and know how to apply principal component sin process monitoring
12. Understand the basis of profile monitoring

Important Terms and Concepts

Average run length
Chi-square control chart
Covariance matrix
Matrix of scatter plots
Monitoring multivariate variability
Multivariate normal distribution
Partial least squares
Phase II control limits
Principal components analysis (PCA)
Profiles
Residual control chart
Sample mean vector
Trajectory plots

Cascade process
Control ellipse
Hotelling T^2 control chart
Mean vector
Multivariate EWMA control chart
Multivariate process control
Phase I control limits
Principal component scores
Profile monitoring
Regression adjustment
Sample covariance matrix
Subgroup data versus individual observations

Exercises

Note: MINITAB's **Tsquared** functionality does not use summary statistics, so many of these exercises were solved in Excel.

10-1.   The data shown here come from a production process with two observable quality characteristics, x_1 and x_2. The data are sample means of each quality characteristic, based on samples of size $n = 25$. Assume that mean values of the quality characteristics and the covariance matrix were computed from 50 preliminary samples:

$$\bar{x} = \begin{bmatrix} 55 \\ 30 \end{bmatrix} \qquad S = \begin{bmatrix} 200 & 130 \\ 130 & 120 \end{bmatrix}$$

Construct a T^2 control chart using these data. Use the phase II limits.

Sample Number	$\bar{x}_1$	$\bar{x}_2$
1	58	32
2	60	33
3	50	27
4	54	31
5	63	38
6	53	30
7	42	20
8	55	31
9	46	25
10	50	29
11	49	27
12	57	30
13	58	33
14	75	45
15	55	27

Phase II T^2 control charts with $m = 50$ preliminary samples, $n = 25$ sample size, $p = 2$ characteristics. Let $\alpha = 0.001$.

$$UCL = \frac{p(m+1)(n-1)}{mn-m-p+1} F_{\alpha, p, mn-m-p+1}$$

$$= \frac{2(50+1)(25-1)}{50(25)-50-2+1} F_{0.001,2,1199} \qquad \text{(Equation 10-21)}$$

$$= (2448/1199)(6.948) = 14.186$$

$$LCL = 0$$

10-1 continued

This exercise may be solved in Excel.

Formulas:

	A	B	C
1	p =	2	xdoublebars
2	m =	50	55
3	n =	25	30
4	alpha =	0.001	
5	F =	=FINV(B4,B1,B2*B3-B2-B1+1)	
6			
7	Sample No.	1	2
8	xbar1	58	60
9	xbar2	32	33
10			
11	diff1	=B8-$C$2	=C8-$C$2
12	diff2	=B9-$C$3	=C9-$C$3
13			
14	matrix calc	=MMULT(MMULT(TRANSPOSE(B11:B12),$F$2:$G$3),B11:B12)	=MMULT(MMULT(TRANSPOSE(C11:C12),$F$2:$G$3),C11:C12)
15	t2 = n * calc	=25*B14	=25*C14
16			
17	UCL =	=($B$1*($B$2+1)*($B$3-1))/(($B$2*$B$3)-$B$2-$B$1+1)*$B$5	=($B$1*($B$2+1)*($B$3-1))/(($B$2*$B$3)-$B$2-$B$1+1)*$B$5
18	LCL =	0	0
19			
20	OOC?	=IF(B15>B17,"Above UCL",IF(B15<B18,"Below LCL","In control"))	=IF(C15>C17,"Above UCL",IF(C15<C18,"Below LCL","In control"))

	D	E
1	S: Var-Covar matrix	
2	200	130
3	130	120
4		
5		
6		
7	3	4
8	50	54
9	27	31
10		
11	=D8-$C$2	=E8-$C$2
12	=D9-$C$3	=E9-$C$3
13		
14	=MMULT(MMULT(TRANSPOSE(D11:D12),$F$2:$G$3),D11:D12)	=MMULT(MMULT(TRANSPOSE(E11:E12),$F$2:$G$3),E11:E12)
15	=25*D14	=25*E14
16		
17	=($B$1*($B$2+1)*($B$3-1))/(($B$2*$B$3)-$B$2-$B$1+1)*$B$5	=($B$1*($B$2+1)*($B$3-1))/(($B$2*$B$3)-$B$2-$B$1+1)*$B$5
18	0	0
19		
20	=IF(D15>D17,"Above UCL",IF(D15<D18,"Below LCL","In control"))	=IF(E15>E17,"Above UCL",IF(E15<E18,"Below LCL","In control"))

	F	G
1	S-1	
2	=MINVERSE(D2:E3)	=MINVERSE(D2:E3)
3	=MINVERSE(D2:E3)	=MINVERSE(D2:E3)
4		
5		
6		
7	5	6
8	63	53
9	38	30
10		
11	=F8-$C$2	=G8-$C$2
12	=F9-$C$3	=G9-$C$3
13		
14	=MMULT(MMULT(TRANSPOSE(F11:F12),$F$2:$G$3),F11:F12)	=MMULT(MMULT(TRANSPOSE(G11:G12),$F$2:$G$3),G11:G12)
15	=25*F14	=25*G14
16		
17	=($B$1*($B$2+1)*($B$3-1))/(($B$2*$B$3)-$B$2-$B$1+1)*$B$5	=($B$1*($B$2+1)*($B$3-1))/(($B$2*$B$3)-$B$2-$B$1+1)*$B$5
18	0	0
19		
20	=IF(F15>F17,"Above UCL",IF(F15<F18,"Below LCL","In control"))	=IF(G15>G17,"Above UCL",IF(G15<G18,"Below LCL","In control"))

10-1 continued

Results:

	A	B	C	D	E	F	G	H	I
1	p =	2	xdoublebars	S: Var-Covar matrix		S-1			
2	m =	50	55	200	130	0.0169	-0.0183		
3	n =	25	30	130	120	-0.0183	0.0282		
4	alpha =	0.001							
5	F =	6.9476							
6									
7	Sample No.	1	2	3	4	5	6	7	8
8	xbar1	58	60	50	54	63	53	42	55
9	xbar2	32	33	27	31	38	30	20	31
10									
11	diff1	3	5	-5	-1	8	-2	-13	0
12	diff2	2	3	-3	1	8	0	-10	1
13									
14	matrix calc	0.0451	0.1268	0.1268	0.0817	0.5408	0.0676	0.9127	0.0282
15	t2 = n * calc	1.1268	3.1690	3.1690	2.0423	13.5211	1.6901	22.8169	0.7042
16									
17	UCL =	14.1850	14.1850	14.1850	14.1850	14.1850	14.1850	14.1850	14.1850
18	LCL =	0	0	0	0	0	0	0	0
19									
20	OOC?	In control	In control	In control	In control	In control	In control	Above UCL	In control

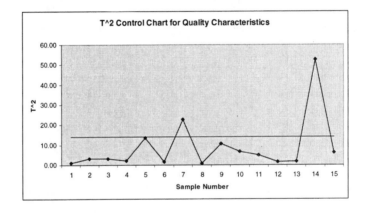

T^2 Control Chart for Quality Characteristics

Process is out of control at samples 7 and 14.

10-3. Reconsider the situation in Exercise 10-1. Suppose that the sample mean vector and sample covariance matrix provided were the actual population parameters. What control limit would be appropriate for phase II for the control chart? Apply this limit to the data and discuss any differences in results that you find in comparison to the original choice of control limit.

Phase II T^2 control limits with $p = 2$ characteristics. Let $\alpha = 0.001$.
Since population parameters are known, the chi-square formula will be used for the upper control limit:
$$\text{UCL} = \chi^2_{\alpha,p} = \chi^2_{0.001,2} = 13.816 \text{ (Equation 10-13)}$$

This exercise may be solved in Excel.

Results:

	A	B	C	D	E	F	G	H	I	J	K	L	M	N	O	P
1	p =	2		xdoublebars		S: Var-Covar matrix			S-1							
2	m =	50		55		200	130		0.0169	-0.0183						
3	n =	25		30		130	120		-0.0183	0.0282						
4	alpha =	0.001														
5	X^2 =	13.8150														
6																
7	Sample No.	1	2	3	4	5	6	7	8	9	10	11	12	13	14	15
8	xbar1	58	60	50	54	63	53	42	55	46	50	49	57	58	75	55
9	xbar2	32	33	27	31	38	30	20	31	25	29	27	30	33	45	27
10																
11	diff1	3	5	-5	-1	8	-2	-13	0	-9	-5	-6	2	3	20	0
12	diff2	2	3	-3	1	8	0	-10	1	-5	-1	-3	0	3	15	-3
13																
14	matrix calc	0.0451	0.1268	0.1268	0.0817	0.5408	0.0676	0.9127	0.0282	0.4254	0.2676	0.2028	0.0676	0.0761	2.1127	0.2535
15	t2 = n * calc	1.1268	3.1690	3.1690	2.0423	13.5211	1.6901	22.8169	0.7042	10.6338	6.6901	5.0704	1.6901	1.9014	52.8169	6.3380
16																
17	UCL =	13.8150	13.8150	13.8150	13.8150	13.8150	13.8150	13.8150	13.8150	13.8150	13.8150	13.8150	13.8150	13.8150	13.8150	13.8150
18	LCL =	0	0	0	0	0	0	0	0	0	0	0	0	0	0	0
19																
20	OOC?	In control	In control	In control	In control	In control	In control	Above UCL	In control	In control	In control	In control	In control	In control	Above UCL	In control

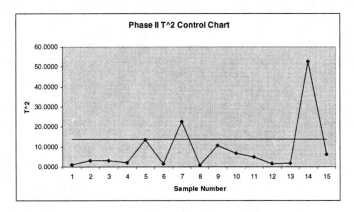

Process is out of control at samples 7 and 14. Same results as for parameters estimated from samples.

10-5. Consider a T^2 control chart for monitoring $p = 6$ quality characteristics. Suppose that the subgroup size is $n = 3$ and there are 30 preliminary samples available to estimate the sample covariance matrix.

(a) Find the phase II control limits assuming that $\alpha = 0.005$.

$m = 30$ preliminary samples, $n = 3$ sample size, $p = 6$ characteristics, $\alpha = 0.005$

Phase II limits

$$UCL = \frac{p(m+1)(n-1)}{mn-m-p+1} F_{\alpha,p,mn-m-p+1} = \frac{6(30+1)(3-1)}{30(3)-30-6+1} F_{0.005,6,55} = \left(\frac{372}{55}\right)(3.531) = 23.882 \text{ (Equation 10-21)}$$

$LCL = 0$

(b) Compare the control limits from part (a) to the chi-square control limit. What is the magnitude of the difference in the two control limits?

chi-square limit: $UCL = \chi^2_{\alpha,p} = \chi^2_{0.005,6} = 18.548$ (Equation 10-13)

The Phase II UCL is almost 30% larger than the chi-square limit.

(c) How many preliminary samples would have to be taken to ensure that the exact phase II control limit is within 1% of the chi-square control limit?

Quality characteristics, $p = 6$. Samples size, $n = 3$. $\alpha = 0.005$. Find "m" such that exact Phase II limit is within 1% of chi-square limit, $1.01(18.548) = 18.733$. This exercise may be solved in Excel.

Formulas:

	A	B	C	D	E
1	p =	6	quality characteristics		
2	n =	3	sample size		
3	alpha =	0.005			
4	Find "m" su				
5					
6	m	num	denom	F	UCL
7	30	=$B$1*(A7+1)*($B$2-1)	=A7*$B$2-A7-$B$1+1	=FINV($B$3,$B$1,C7)	=B7/C7*D7
8	=A7+10	=$B$1*(A8+1)*($B$2-1)	=A8*$B$2-A8-$B$1+1	=FINV($B$3,$B$1,C8)	=B8/C8*D8
9	=A8+10	=$B$1*(A9+1)*($B$2-1)	=A9*$B$2-A9-$B$1+1	=FINV($B$3,$B$1,C9)	=B9/C9*D9
10	=A9+10	=$B$1*(A10+1)*($B$2-1)	=A10*$B$2-A10-$B$1+1	=FINV($B$3,$B$1,C10)	=B10/C10*D10
11	=A10+10	=$B$1*(A11+1)*($B$2-1)	=A11*$B$2-A11-$B$1+1	=FINV($B$3,$B$1,C11)	=B11/C11*D11
12	...				

Results:

	A	B	C	D	E
6	m	num	denom	F	UCL
7	30	372		55	3.531	23.8820
8	40	492		75	3.407	22.3527
9	50	612		95	3.338	21.5042
10	60	732		115	3.294	20.9650
11	70	852		135	3.263	20.5920
83	...				
84	718	8628		1431	3.107	18.7332
85	719	8640		1433	3.107	18.7331
86	720	8652		1435	3.107	18.7328
87	721	8664		1437	3.107	18.7325
88	722	8676		1439	3.107	18.7324

720 preliminary samples must be taken to ensure that the exact Phase II limit is within 1% of the chi-square limit.

10-7.   Consider a T^2 control chart for monitoring $p = 10$ quality characteristics. Suppose that the subgroup size is $n = 3$ and there are 25 preliminary samples available to estimate the sample covariance matrix.

(a) Find the phase II control limits assuming that $\alpha = 0.005$.

   $m = 25$ preliminary samples, $n = 3$ sample size, $p = 10$ characteristics, $\alpha = 0.005$

   Phase II UCL

$$UCL = \frac{p(m+1)(n-1)}{mn-m-p+1} F_{\alpha,p,mn-m-p+1} = \frac{10(25+1)(3-1)}{25(3)-25-10+1} F_{0.005,10,41} = \left(\frac{520}{41}\right)(3.101) = 39.326 \ \ (\text{Eqn 10-21})$$

(b) Compare the control limits from part (a) to the chi-square control limit. What is the magnitude of the difference in the two control limits?

   chi-square UCL: $UCL = \chi^2_{\alpha,p} = \chi^2_{0.005,10} = 25.188$ (Equation 10-13)

   The Phase II UCL is more than 55% larger than the chi-square limit.

(c) How many preliminary samples would have to be taken to ensure that the chi-square control limit is within 1% of the exact phase II control limit?

   Quality characteristics, $p = 10$. Samples size, $n = 3$. $\alpha = 0.005$. Find "m" such that exact Phase II limit is within 1% of chi-square limit, $1.01(25.188) = 25.440$. This exercise may be solved in Excel.

Formulas:

	A	B	C	D	E
1	p =	10	quality characteristics		
2	n =	3	sample size		
3	alpha =	0.005			
4	Find "m" s				
5					
6	m	num	denom	F	UCL
7	25	=$B$1*(A7+1)*($B$2-1)	=A7*$B$2-A7-$B$1+1	=FINV($B$3,$B$1,C7)	=B7/C7*D7
8	=A7+10	=$B$1*(A8+1)*($B$2-1)·	=A8*$B$2-A8-$B$1+1	=FINV($B$3,$B$1,C8)	=B8/C8*D8
9	=A8+10	=$B$1*(A9+1)*($B$2-1)	=A9*$B$2-A9-$B$1+1	=FINV($B$3,$B$1,C9)	=B9/C9*D9
10	=A9+10	=$B$1*(A10+1)*($B$2-1)	=A10*$B$2-A10-$B$1+1	=FINV($B$3,$B$1,C10)	=B10/C10*D10
11	=A10+10	=$B$1*(A11+1)*($B$2-1)	=A11*$B$2-A11-$B$1+1	=FINV($B$3,$B$1,C11)	=B11/C11*D11
12	...				

Results:

	A	B	C	D	E
6	m	num	denom	F	UCL
7	25	520		41	3.101	39.3259
8	35	720		61	2.897	34.1991
9	45	920		81	2.799	31.7953
10	55	1120		101	2.742	30.4024
11	65	1320		121	2.704	29.4940
12	...				
104	985	19720		1961	2.530	25.4406
105	986	19740		1963	2.530	25.4405
106	987	19760		1965	2.530	25.4401
107	988	19780		1967	2.530	25.4399
108	989	19800		1969	2.530	25.4398
109	990	19820		1971	2.530	25.4394

988 preliminary samples must be taken to ensure that the exact Phase II limit is within 1% of the chi-square limit.

10-9.   Consider a T^2 control chart for monitoring $p = 10$ quality characteristics. Suppose that the subgroup size is $n = 3$ and there are 25 preliminary samples available to estimate the sample covariance matrix. Calculate both the phase I and the phase II control limits (use $\alpha = 0.01$).

$p = 10$ quality characteristics, $n = 3$ sample size, $m = 25$ preliminary samples. Assume $\alpha = 0.01$.

Phase I UCL

$$\begin{aligned}
\text{UCL} &= \frac{p(m-1)(n-1)}{mn-m-p+1} F_{\alpha,p,mn-m-p+1} \\
&= \frac{10(25-1)(3-1)}{25(3)-25-10+1} F_{0.01,10,41} \\
&= \left(\frac{480}{41}\right)(2.788) \\
&= 32.638
\end{aligned}$$

Phase II UCL

$$\begin{aligned}
\text{UCL} &= \frac{p(m+1)(n-1)}{mn-m-p+1} F_{\alpha,p,mn-m-p+1} \\
&= \frac{10(25+1)(3-1)}{25(3)-25-10+1} F_{0.01,10,41} \\
&= \left(\frac{520}{41}\right)(2.788) \\
&= 35.360
\end{aligned}$$

10-11.   Suppose that we have $p = 3$ quality characteristics, and in correlation form all three variables have variance unity and all pairwise correlation coefficients are 0.8. The in-control value of the process mean vector is $\mu' = [0, 0, 0]$.

(a) Write out the covariance matrix Σ.

$$\Sigma = \begin{bmatrix} 1 & 0.8 & 0.8 \\ 0.8 & 1 & 0.8 \\ 0.8 & 0.8 & 1 \end{bmatrix}$$

(b) What is the chi-square control limit for the chart, assuming that $\alpha = 0.05$?

$$\text{UCL} = \chi^2_{\alpha,p} = \chi^2_{0.05,3} = 7.815$$

(c) Suppose that a sample of observations results in the standardized observation vector $\mathbf{y}' = [1, 2, 0]$. Calculate the value of the T^2 statistic. Is an out-of-control signal generated?

$$T^2 = 11.154$$

Yes. Since $(T^2 = 11.154) > (\text{UCL} = 7.815)$, an out-of-control signal is generated.

(d) Calculate the diagnostic quantities d_i, $i = 1, 2, 3$ from equation 10-22. Does this information assist in identifying which process variables have shifted?

$$\chi^2_{0.05,1} = 3.841$$

$$T^2_{(1)} = 11.111; \quad d_1 = 0.043$$

$$T^2_{(2)} = 2.778; \quad d_2 = 8.376$$

$$T^2_{(3)} = 5.000; \quad d_3 = 6.154$$

Variables 2 and 3 should be investigated.

(e) Suppose that a sample of observations results in the standardized observation vector $\mathbf{y}' = [2, 2, 1]$. Calculate the value of the T^2 statistic. Is an out-of-control signal generated?

Since $(T^2 = 6.538) > (\text{UCL} = 7.815)$, an out-of-control signal is not generated.

(f) For the case in (e), calculate the diagnostic quantities d_i, $i = 1, 2, 3$ from equation 10-22. Does this information assist in identifying which process variables have shifted?

$$\chi^2_{0.05,1} = 3.841$$

$$T^2_{(1)} = 5.000; \quad d_1 = 1.538$$

$$T^2_{(2)} = 5.000; \quad d_2 = 1.538$$

$$T^2_{(3)} = 4.444; \quad d_3 = 2.094$$

Since an out-of-control signal was not generated in (e), it is not necessary to calculate the diagnostic quantities. This is confirmed since none of the d_i's exceeds the UCL.

10-13. Consider the first three process variables in Table 10-5. Calculate an estimate of the sample covariance matrix using both estimators S_1 and S_2 discussed in Section 10-3.2.

Table 10-5 Cascade Process Data

Observation	x_1	x_2	x_3					
1	12.78	0.15	91	21	15.26	0.13	91	47
2	14.97	0.1	90	22	17.3	0.12	95	47
3	15.43	0.07	90	23	17.62	0.06	95	42
4	14.95	0.12	89	24	18.21	0.06	93	41
5	16.17	0.1	83	25	14.38	0.1	90	46
6	17.25	0.07	84	26	12.13	0.14	87	50
7	16.57	0.12	89	27	12.72	0.1	90	47
8	19.31	0.08	99	28	17.42	0.1	89	51
9	18.75	0.04	99	29	17.63	0.11	87	45
10	16.99	0.09	98	30	16.17	0.05	83	57
11	18.2	0.13	98	31	16.88	0.16	86	58
12	16.2	0.16	97	32	13.87	0.16	85	46
13	14.72	0.12	82	33	14.56	0.05	84	41
14	14.42	0.13	81	34	15.35	0.12	83	40
15	11.02	0.1	83	35	15.91	0.12	81	45
16	9.82	0.1	86	36	14.32	0.11	85	47
17	11.41	0.12	87	37	15.43	0.13	86	43
18	14.74	0.1	81	38	14.47	0.08	85	54
19	14.5	0.08	84	39	14.74	0.07	84	52
20	14.71	0.09	89	40	16.28	0.13	86	49

This exercise may be solved in Excel.

$m = 40$

$$\bar{x}' = \begin{bmatrix} 15.339 & 0.104 & 88.125 \end{bmatrix}; \quad S_1 = \begin{bmatrix} 4.440 & -0.016 & 5.395 \\ -0.016 & 0.001 & -0.014 \\ 5.395 & -0.014 & 27.599 \end{bmatrix}$$

$$V'V = \begin{bmatrix} 121.101 & -0.256 & 43.720 \\ -0.256 & 0.071 & 0.950 \\ 43.720 & 0.950 & 587.000 \end{bmatrix}; \quad S_2 = \begin{bmatrix} 1.553 & -0.003 & -0.561 \\ -0.003 & 0.001 & 0.012 \\ -0.561 & 0.012 & 7.526 \end{bmatrix}$$

	A	B	C	D	E	F	G	I	J	K	M	N	O
1	x1	x2	x3		x1-xbar1	x2-xbar2	x3-xbar3	(x1-xb1)*(x2-xb2)	(x1-xb1)*(x3-xb3)	(x2-xb2)*(x3-xb3)	v1	v2	v3
2	12.78	0.15	91		-2.559	0.046	2.875	-0.117	-7.357	0.132	2.190	-0.050	-1.000
3	14.97	0.1	90		-0.369	-0.004	1.875	0.002	-0.692	-0.008	0.460	-0.030	0.000
4	15.43	0.07	90		0.091	-0.034	1.875	-0.003	0.171	-0.064	-0.480	0.050	-1.000
5	14.95	0.12	89		-0.389	0.016	0.875	-0.006	-0.340	0.014	1.220	-0.020	-6.000
6	16.17	0.1	83		0.831	-0.004	-5.125	-0.004	-4.259	0.022	1.080	-0.030	1.000
7	...												

10-13 continued

	Q	R	S	T	U	V	W
1	xbar						
2	xbar1	=AVERAGE(A2:A41)					
3	xbar2	=AVERAGE(B2:B41)					
4	xbar3	=AVERAGE(C2:C41)					
5							
6	S1						
7	=VAR(A2:A41)	-0.0162803	5.39474358974359				
8	=SUM(I2:I41)/(40-1)	=VAR(B2:B41)	-0.0141346153846154				
9	=SUM(J2:J41)/(40-1)	=SUM(K2:K41)/(40-1)	=VAR(C2:C41)				
10							
11	V'V				S2		
12	=MMULT(R16:BD18,M2:O40)	=MMULT(R16:BD18,M2:O40)	=MMULT(R16:BD18,M2:O40)		=Q12/(2*(40-1))	=U13	=U14
13	=MMULT(R16:BD18,M2:O40)	=MMULT(R16:BD18,M2:O40)	=MMULT(R16:BD18,M2:O40)		=Q13/(2*(40-1))	=R13/(2*(40-1))	=V14
14	=MMULT(R16:BD18,M2:O40)	=MMULT(R16:BD18,M2:O40)	=MMULT(R16:BD18,M2:O40)		=Q14/(2*(40-1))	=R14/(2*(40-1))	=S14/(2*(40-1))
15							
16	v1'	2.19	0.459999999999999	-0.48	1.22	1.08	-0.68
17	v2'	-0.05	-0.03	0.05	-0.02	-0.03	0.05
18	v3'	-1	0	-1	-6	1	5

	Q	R	S	T	U	V	W
1	xbar						
2	xbar1	15.339					
3	xbar2	0.104					
4	xbar3	88.125					
5							
6	S1						
7		4.440	-0.016	5.395			
8		-0.016	0.001	-0.014			
9		5.395	-0.014	27.599			
10							
11	V'V				S2		
12		121.101	-0.256	43.720		1.553	-0.003	0.561
13		-0.256	0.071	0.950		-0.003	0.001	0.012
14		43.720	0.950	587.000		0.561	0.012	7.526
15							
16	v1'	2.190	0.460	-0.480	1.220	1.080	-0.680
17	v2'	-0.050	-0.030	0.050	-0.020	-0.030	0.050
18	v3'	-1.000	0.000	-1.000	-6.000	1.000	5.000

10-15. Suppose that there are $p = 4$ quality characteristics, and in correlation form all four variables have variance unity and all pairwise correlation coefficients are 0.75. The in-control value of the process mean vector is $\mu' = [0, 0, 0, 0]$, and we want to design an MEWMA control chart to provide good protection against a shift to a new mean vector of $y' = [1, 1, 1, 1]$. If an in-control ARL_0 of 200 is satisfactory, what value of λ and what upper control limit should be used? Approximately, what is the ARL_1 for detecting the shift in the mean vector?

This exercise may be solved in Excel.

	A	B	C	D	E
1	p =	4			
2	mu' =	0	0	0	0
3					
4	Sigma =	1	0.75	0.75	0.75
5		0.75	1	0.75	0.75
6		0.75	0.75	1	0.75
7		0.75	0.75	0.75	1
8					
9	Sigma-1 =	=MINVERSE(B4:E7)	=MINVERSE(B4:E7)	=MINVERSE(B4:E7)	=MINVERSE(B4:E7)
10		=MINVERSE(B4:E7)	=MINVERSE(B4:E7)	=MINVERSE(B4:E7)	=MINVERSE(B4:E7)
11		=MINVERSE(B4:E7)	=MINVERSE(B4:E7)	=MINVERSE(B4:E7)	=MINVERSE(B4:E7)
12		=MINVERSE(B4:E7)	=MINVERSE(B4:E7)	=MINVERSE(B4:E7)	=MINVERSE(B4:E7)
13					
14	y' =	=B16	=B17	=B18	=B19
15					
16	y =	1			
17		1			
18		1			
19		1			
20					
21	y' Sigma-1 =	=MMULT(B14:E14,B9:E12)	=MMULT(B14:E14,B9:E12)	=MMULT(B14:E14,B9:E12)	=MMULT(B14:E14,B9:E12)
22					
23	y' Sigma-1 y =	=MMULT(B21:E21,B16:B19)			
24					
25	delta =	=SQRT(B23)			
26					
27	ARL0 =	200			
28					
29	From Table 10-3				
30	delta =	1	1.5		
31	lambda =	0.1	0.2		
32	UCL = H =	12.73	13.87		
33	ARL1 =	12.17	6.53		

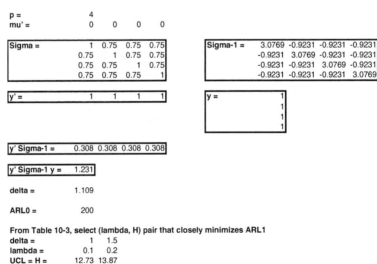

p =　　　4
mu' =　　0　0　0　0

Sigma =	1	0.75	0.75	0.75
	0.75	1	0.75	0.75
	0.75	0.75	1	0.75
	0.75	0.75	0.75	1

Sigma-1 =	3.0769	-0.9231	-0.9231	-0.9231
	-0.9231	3.0769	-0.9231	-0.9231
	-0.9231	-0.9231	3.0769	-0.9231
	-0.9231	-0.9231	-0.9231	3.0769

y' =	1	1	1	1

y =	1
	1
	1
	1

y' Sigma-1 =	0.308	0.308	0.308	0.308

y' Sigma-1 y =	1.231

delta =　　1.109

ARL0 =　　200

From Table 10-3, select (lambda, H) pair that closely minimizes ARL1
delta =　　1　1.5
lambda =　0.1　0.2
UCL = H =　12.73　13.87
ARL1 =　　12.17　6.53

Select $\lambda = 0.1$ with an UCL = $H = 12.73$. This gives an ARL_1 between 7.22 and 12.17.

10-17. Suppose that there are $p = 2$ quality characteristics, and in correlation form both variables have variance unity and the correlation coefficient is 0.8. The in-control value of the process mean vector is $\mu' = [0, 0]$, and we want to design an MEWMA control chart to provide good protection against a shift to a new mean vector of $y' = [1, 1]$. If an in-control ARL_0 of 200 is satisfactory, what value of λ and what upper control limit should be used? Approximately, what is the ARL_1 for detecting the shift in the mean vector?

This exercise may be solved in Excel.

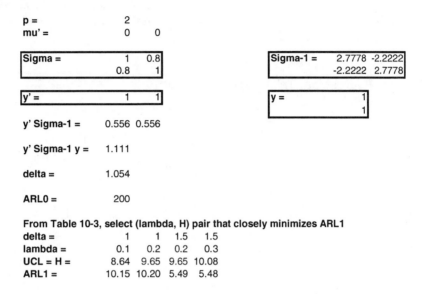

p =	2		
mu' =	0	0	

Sigma =	1	0.8		Sigma-1 =	2.7778	-2.2222
	0.8	1			-2.2222	2.7778

y' =	1	1		y =	1
					1

y' Sigma-1 =   0.556 0.556

y' Sigma-1 y =   1.111

delta =   1.054

ARL0 =   200

From Table 10-3, select (lambda, H) pair that closely minimizes ARL1

delta =	1	1	1.5	1.5
lambda =	0.1	0.2	0.2	0.3
UCL = H =	8.64	9.65	9.65	10.08
ARL1 =	10.15	10.20	5.49	5.48

Select $\lambda = 0.2$ with an UCL = $H = 9.65$. This gives an ARL_1 between 5.49 and 10.20.

10-19.   Consider the cascade process data in Table 10-5. In fitting regression models to both y_1 and y_2 you will find that not all of the process variables are required to obtain a satisfactory regression model for the output variables. Remove the nonsignificant variables from these equations and obtain subset regression models for both y_1 and y_2. Then construct individuals control charts for both sets of residuals. Compare them to the residual control charts in the text (Fig. 10-11) and from Exercise 10-18. Are there any substantial differences between the charts from the two different approaches to fitting the regression models?

Different approaches can be used to identify insignificant variables and reduce the number of variables in a regression model. This solution uses MINITAB's "Best Subsets" functionality to identify the best-fitting model with as few variables as possible.

Stat > Regression > Best Subsets

Best Subsets Regression: Tab10-5y1 versus Tab10-5x1, Tab10-5x2, ...

```
Response is Tab10-5y1
                                          T T T T T T T T T
                                          a a a a a a a a a
                                          b b b b b b b b b
                                          1 1 1 1 1 1 1 1 1
                                          0 0 0 0 0 0 0 0 0
                                          - - - - - - - - -
                                          5 5 5 5 5 5 5 5 5
                           Mallows         x x x x x x x x x
Vars   R-Sq   R-Sq(adj)     C-p      S     1 2 3 4 5 6 7 8 9
   1   43.1      41.6      52.9   1.3087                 X
   1   31.3      29.5      71.3   1.4378       X
   2   62.6      60.5      24.5   1.0760   X             X
   2   55.0      52.5      36.4   1.1799       X         X
   3   67.5      64.7      18.9   1.0171   X   X         X
   3   66.8      64.0      19.9   1.0273   X           X X
   4   72.3      69.1      13.3   0.95201  X   X X       X
   4   72.1      68.9      13.6   0.95522  X   X       X X
   5   79.5      76.5       4.0   0.83020  X   X X     X X    ******
   5   73.8      69.9      13.0   0.93966  X   X     X X X
   6   79.9      76.2       5.5   0.83550  X   X X   X X X
   6   79.8      76.1       5.6   0.83693  X X X X     X X
   7   80.3      76.0       6.8   0.83914  X   X X X X X X
   7   80.1      75.8       7.1   0.84292  X   X X X   X X
...
```

10-19 continued

```
Best Subsets Regression: Tab10-5y2 versus Tab10-5x1, Tab10-5x2, ...
Response is Tab10-5y2
                                           T T T T T T T T T
                                           a a a a a a a a a
                                           b b b b b b b b b
                                           1 1 1 1 1 1 1 1 1
                                           0 0 0 0 0 0 0 0 0
                                           - - - - - - - - -
                                           5 5 5 5 5 5 5 5 5
                               Mallows      x x x x x x x x x
   Vars  R-Sq  R-Sq(adj)        C-p     S   1 2 3 4 5 6 7 8 9
      1  36.1      34.4        24.0  4.6816      X
      1  35.8      34.1        24.2  4.6921                    X
      2  55.1      52.7         8.1  3.9751                  X X
      2  50.7      48.1        12.2  4.1665      X           X
      3  61.6      58.4         4.0  3.7288      X X         X
      3  59.8      56.4         5.7  3.8160      X           X X
      4  64.9      60.9         2.9  3.6147  X   X           X X
      4  64.4      60.4         3.4  3.6387      X X         X X
      5  67.7      62.9         2.3  3.5208  X   X X         X X    ******
      5  65.2      60.1         4.7  3.6526  X   X       X   X X
      6  67.8      62.0         4.2  3.5660  X   X X       X X X
      6  67.8      61.9         4.3  3.5684  X   X X X       X X
      7  67.9      60.9         6.1  3.6149  X X X X       X X X
      7  67.8      60.8         6.2  3.6200  X   X X X       X X X
...
```

For output variables $y1$ and $y2$, a regression model of input variables $x1$, $x3$, $x4$, $x8$, and $x9$ maximize adjusted R^2 (minimize S) and minimize Mallow's C-p.

Stat > Regression > Regression

```
Regression Analysis: Tab10-5y1 versus Tab10-5x1, Tab10-5x3, ...
The regression equation is
Tab10-5y1 = 819 + 0.431 Tab10-5x1 - 0.124 Tab10-5x3 - 0.0915 Tab10-5x4
            + 2.64 Tab10-5x8 + 115 Tab10-5x9

Predictor       Coef   SE Coef       T       P
Constant      818.80     29.14   28.10   0.000
Tab10-5x1    0.43080   0.08113    5.31   0.000
Tab10-5x3   -0.12396   0.03530   -3.51   0.001
Tab10-5x4   -0.09146   0.02438   -3.75   0.001
Tab10-5x8     2.6367    0.7604    3.47   0.001
Tab10-5x9     114.81     23.65    4.85   0.000

S = 0.830201   R-Sq = 79.5%   R-Sq(adj) = 76.5%

Analysis of Variance
Source           DF        SS       MS       F       P
Regression        5    90.990   18.198   26.40   0.000
Residual Error   34    23.434    0.689
Total            39   114.424
```

10-19 continued

Regression Analysis: Tab10-5y2 versus Tab10-5x1, Tab10-5x3, ...

```
The regression equation is
Tab10-5y2 = 244 - 0.633 Tab10-5x1 + 0.454 Tab10-5x3 + 0.176 Tab10-5x4
            + 11.2 Tab10-5x8 - 236 Tab10-5x9


Predictor       Coef  SE Coef       T      P
Constant       244.4    123.6    1.98  0.056
Tab10-5x1    -0.6329   0.3441   -1.84  0.075
Tab10-5x3     0.4540   0.1497    3.03  0.005
Tab10-5x4     0.1758   0.1034    1.70  0.098
Tab10-5x8     11.175    3.225    3.47  0.001
Tab10-5x9     -235.7    100.3   -2.35  0.025


S = 3.52081   R-Sq = 67.7%   R-Sq(adj) = 62.9%


Analysis of Variance
Source          DF       SS       MS      F      P
Regression       5   882.03   176.41  14.23  0.000
Residual Error  34   421.47    12.40
Total           39  1303.50
```

Save residuals from both models to evaluate process control.

Stat > Control Charts > Variables Charts for Individuals > Individuals

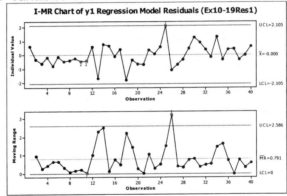

Test Results for I Chart of Ex10-19Res1

```
TEST 1. One point more than 3.00 standard deviations from center line.
Test Failed at points:   25
TEST 2. 9 points in a row on same side of center line.
Test Failed at points:   10, 11
```

Test Results for MR Chart of Ex10-19Res1

```
TEST 1. One point more than 3.00 standard deviations from center line.
Test Failed at points:   26
TEST 2. 9 points in a row on same side of center line.
Test Failed at points:   11
```

10-19 continued

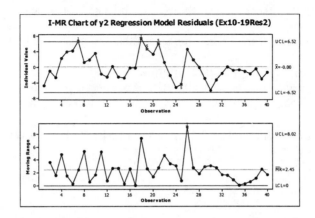

Test Results for I Chart of Ex10-19Res2

```
TEST 1. One point more than 3.00 standard deviations from center line.
Test Failed at points:  7, 18
TEST 5. 2 out of 3 points more than 2 standard deviations from center line (on
     one side of CL).
Test Failed at points:  19, 21, 25
TEST 6. 4 out of 5 points more than 1 standard deviation from center line (on
     one side of CL).
Test Failed at points:  7, 21
```

Test Results for MR Chart of Ex10-19Res2

```
TEST 1. One point more than 3.00 standard deviations from center line.
Test Failed at points:  26
```

For response $y1$, there is not a significant difference between control charts for residuals from either the full regression model (Figure 10-10, no out-of-control observations) and the subset regression model (observation 25 is OOC).

For response $y2$, there is not a significant difference between control charts for residuals from either the full regression model (Exercise 10-18, observations 7 and 18 are OOC) and the subset regression model (observations 7 and 18 are OOC).

10-21.   Consider the $p = 4$ process variables in Table 10-6. After applying the PCA procedure to the first 20 observations data (see Table 10-7), suppose that the first three principal components are retained.

(a) Obtain the principal component scores. (Hint: Remember that you must work in standardized variables.)

Stat > Multivariate > Principal Components
Note: To work in standardized variables in MINITAB, select Correlation Matrix.
Note: To obtain principal component scores, select Storage and enter columns for Scores.

Principal Component Analysis: Ex10-21X1, Ex10-21X2, Ex10-21X3, Ex10-21X4
Eigenanalysis of the Correlation Matrix

Eigenvalue	2.3181	1.0118	**0.6088**	0.0613
Proportion	0.580	0.253	**0.152**	0.015
Cumulative	0.580	0.832	**0.985**	1.000

Variable	PC1	PC2	PC3	PC4
Ex10-21X1	0.594	-0.334	0.257	0.685
Ex10-21X2	0.607	-0.330	0.083	-0.718
Ex10-21X3	0.286	0.794	0.534	-0.061
Ex10-21X4	0.444	0.387	-0.801	0.104

Principal Component Scores:

Ex10-21z1	Ex10-21z2	Ex10-21z3
0.29168	-0.60340	0.02496
0.29428	0.49153	1.23823
0.19734	0.64094	-0.20787
0.83902	1.46958	0.03929
3.20488	0.87917	0.12420
0.20327	-2.29514	0.62545
-0.99211	1.67046	-0.58815
-1.70241	-0.36089	1.82157
-0.14246	0.56081	0.23100
-0.99498	-0.31493	0.33164
0.94470	0.50471	0.17976
-1.21950	-0.09129	-1.11787
2.60867	-0.42176	-1.19166
-0.12378	-0.08767	-0.19592
-1.10423	1.47259	0.01299
-0.27825	-0.94763	-1.31445
-2.65608	0.13529	-0.11243
2.36528	-1.30494	0.32286
0.41131	-0.21893	0.64480
-2.14662	-1.17849	-0.86838

10-21 continued

(b) Construct an appropriate set of pairwise plots of the principal component scores.

Graph > Matrix Plot > Simple Matrix of Plots

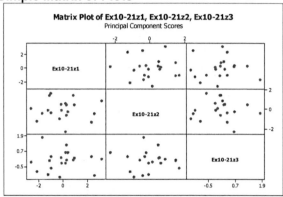

(c) Calculate the principal component scores for the last 10 observations. Plot the scores on the charts from part (b) and interpret the results.

Note: Principal component scores for new observations were calculated in Excel.

	A	B	C	E	F	G	H	I
1	c1	c2	c3	Obs	x1	x2	x3	x4
2	0.594104	-0.333932	0.256985	1	10	20.7	13.6	15.5
3	0.607045	-0.329602	0.083413	2	10.5	19.9	18.1	14.8
4	0.285531	0.793694	0.533676	3	9.7	20	16.1	16.5
5	0.443857	0.387172	-0.801368	4	9.8	20.2	19.1	17.1
6				5	11.7	21.5	19.8	18.3
7				6	11	20.9	10.3	13.8
8				7	8.7	18.8	16.9	16.8
9				8	9.5	19.3	15.3	12.2
10				9	10.1	19.4	16.2	15.8
11				10	9.5	19.6	13.6	14.5
12				11	10.5	20.3	17	16.5
13				12	9.2	19	11.5	16.3
14				13	11.3	21.6	14	18.7
15				14	10	19.8	14	15.9
16				15	8.5	19.2	17.4	15.8
17				16	9.7	20.1	10	16.6
18				17	8.3	18.4	12.5	14.2
19				18	11.9	21.8	14.1	16.2
20				19	10.3	20.5	15.6	15.1
21				20	8.9	19	8.5	14.7
22				mean	=AVERAGE(F2:F21)	=AVERAGE(G2:G21)	=AVERAGE(H2:H21)	=AVERAGE(I2:I21)
23				std dev	=STDEV(F2:F21)	=STDEV(G2:G21)	=STDEV(H2:H21)	=STDEV(I2:I21)
24								
25				21	9.9	20	15.4	15.9
26				22	8.7	19	9.9	16.8
27				23	11.5	21.8	19.3	12.1
28				24	15.9	24.6	14.7	15.3
29				25	12.6	23.9	17.1	14.2
30				26	14.9	25	16.3	16.6
31				27	9.9	23.7	11.9	18.1
32				28	12.8	26.3	13.5	13.7
33				29	13.1	26.1	10.9	16.8
34				30	9.8	25.8	14.8	15

10-21 continued

Calculations for first few scores of z1; score calculations for new observations are similar.

	J
1	**z1**
2	=($F2-$F$22)/$F$23*A$2+($G2-$G$22)/$G$23*A$3+($H2-$H$22)/$H$23*A$4+($I2-$I$22)/$I$23*A$5
3	=($F3-$F$22)/$F$23*A$2+($G3-$G$22)/$G$23*A$3+($H3-H$22)/$H$23*A$4+($I3-$I$22)/$I$23*A$5
4	=($F4-$F$22)/$F$23*A$2+($G4-$G$22)/$G$23*A$3+($H4-H$22)/$H$23*A$4+($I4-$I$22)/$I$23*A$5
5	=($F5-$F$22)/$F$23*A$2+($G5-$G$22)/$G$23*A$3+($H5-H$22)/$H$23*A$4+($I5-$I$22)/$I$23*A$5
6	=($F6-$F$22)/$F$23*A$2+($G6-$G$22)/$G$23*A$3+($H6-H$22)/$H$23*A$4+($I6-$I$22)/$I$23*A$5

Graph > Matrix Plot > Matrix of Plots with Groups

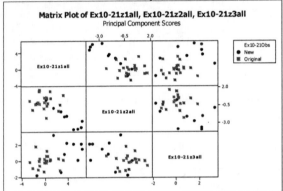

Although a few new points are within area defined by the original points, the majority of new observations are clearly different from the original observations.

CHAPTER 11

Engineering Process Control and SPC

Learning Objectives

After completing this chapter you should be able to:
1. Explain the origins of process monitoring and process adjustment
2. Explain the different statistical frameworks of SPC and EPC
3. Explain how an integral controller works
4. Understand how EPC transfers variability from the process output into a manipulatable variable
5. Know how to set up and use a manual adjustment chart
6. Understand the basis of the bounded adjustment chart
7. Understand the basis of proportional integral (PI) and proportional integral derivative (PID) controllers
8. Know how to combine SPC and EPC by applying a control chart to the output quality characteristic

Important Terms and Concepts

Adjustment chart

Bounded adjustment

Exponentially weighted moving average (EWMA)

Integrating SPC and EPC

Manual adjustment chart

Process gain

Proportional integral derivative control

Statistical process control (SPC)

Automatic process control

Engineering process control (EPC)

Integral control

Manipulatable variable

Process adjustment versus process monitoring

Proportional integral control

Setpoint

Statistical process monitoring

Exercises

Note: MINITAB does not include functionality for some of the methods presented in Chapter 11; however Excel can be used to solve the exercises.

11-1. If y_t are the observations and z_t is the EWMA, show that the following relationships are true.

(a) $z_t - z_t - 1 = \lambda(y_t - z_{t-1})$

   y_t : observation

   z_t : EWMA

$$z_t = \lambda y_t + (1-\lambda)z_{t-1}$$
$$z_t = \lambda y_t + z_{t-1} - \lambda z_{t-1}$$
$$z_t - z_{t-1} = \lambda y_t + z_{t-1} - z_{t-1} - \lambda z_{t-1}$$
$$z_t - z_{t-1} = \lambda y_t - \lambda z_{t-1}$$
$$z_t - z_{t-1} = \lambda(y_t - z_{t-1})$$

(b) $e_t - (1-\lambda)e_t - 1 = y_t - y_{t-1}$

$$z_{t-1} - z_{t-2} = \lambda e_{t-1} \quad \text{(as a result of part (a))}$$
$$z_{t-1} - z_{t-2} + (e_t - e_{t-1}) = \lambda e_{t-1} + (e_t - e_{t-1})$$
$$z_{t-1} + e_t - z_{t-2} - e_{t-1} = e_t - (1-\lambda)e_{t-1}$$
$$y_t - y_{t-1} = e_t - (1-\lambda)e_{t-1}$$

11-3. Consider the data in Table 11-1. Construct a bounded adjustment chart using $\lambda = 0.4$ and $L = 10$. Compare the performance of this chart to the one in Table 11-1 and Fig. 11-12.

This exercise may be solved in Excel.

	A	B	C	D	E
1	Target yt =	0			
2	lambda =	0.4			
3	L =	10			
4	g =	0.8			
5					
6	Obs	Orig_out	Orig_Nt	Adj_out_t	EWMA_t
7	1	0	0		
8	2	16	=B8-B7	=B8	=IF(ABS(E7)<=$B$3,$B$2*D8+(1-$B$2)*E7,$B$2*D8+(1-$B$2)*0)
9	3	24	=B9-B8	=B9	=IF(ABS(E8)<=$B$3,$B$2*D9+(1-$B$2)*E8,$B$2*D9+(1-$B$2)*0)
10	4	29	=B10-B9	=D9+C10+G9	=IF(ABS(E9)<=$B$3,$B$2*D10+(1-$B$2)*E9,$B$2*D10+(1-$B$2)*0)
11	5	34	=B11-B10	=D10+C11+G10	=IF(ABS(E10)<=$B$3,$B$2*D11+(1-$B$2)*E10,$B$2*D11+(1-$B$2)*0)
12	6	24	=B12-B11	=D11+C12+G11	=IF(ABS(E11)<=$B$3,$B$2*D12+(1-$B$2)*E11,$B$2*D12+(1-$B$2)*0)
13	7	31	=B13-B12	=D12+C13+G12	=IF(ABS(E12)<=$B$3,$B$2*D13+(1-$B$2)*E12,$B$2*D13+(1-$B$2)*0)
14	8	26	=B14-B13	=D13+C14+G13	=IF(ABS(E13)<=$B$3,$B$2*D14+(1-$B$2)*E13,$B$2*D14+(1-$B$2)*0)
15	9	38	=B15-B14	=D14+C15+G14	=IF(ABS(E14)<=$B$3,$B$2*D15+(1-$B$2)*E14,$B$2*D15+(1-$B$2)*0)
16	10	29	=B16-B15	=D15+C16+G15	=IF(ABS(E15)<=$B$3,$B$2*D16+(1-$B$2)*E15,$B$2*D16+(1-$B$2)*0)
17	...				

	F	G	H
1			
2			
3			
4			
5			
6	|EWMA_t|>L?	Adj_Obs_t+1	Cum_Adj
7			
8	=IF(ABS(E8)>$B$3,"yes","no")	=IF(ABS(E8)<=$B$3,0,(-$B$2/$B$4*(D8-$B$1)))	=H7+G8
9	=IF(ABS(E9)>$B$3,"yes","no")	=IF(ABS(E9)<=$B$3,0,(-$B$2/$B$4*(D9-$B$1)))	=H8+G9
10	=IF(ABS(E10)>$B$3,"yes","no")	=IF(ABS(E10)<=$B$3,0,(-$B$2/$B$4*(D10-$B$1)))	=H9+G10
11	=IF(ABS(E11)>$B$3,"yes","no")	=IF(ABS(E11)<=$B$3,0,(-$B$2/$B$4*(D11-$B$1)))	=H10+G11
12	=IF(ABS(E12)>$B$3,"yes","no")	=IF(ABS(E12)<=$B$3,0,(-$B$2/$B$4*(D12-$B$1)))	=H11+G12
13	=IF(ABS(E13)>$B$3,"yes","no")	=IF(ABS(E13)<=$B$3,0,(-$B$2/$B$4*(D13-$B$1)))	=H12+G13
14	=IF(ABS(E14)>$B$3,"yes","no")	=IF(ABS(E14)<=$B$3,0,(-$B$2/$B$4*(D14-$B$1)))	=H13+G14
15	=IF(ABS(E15)>$B$3,"yes","no")	=IF(ABS(E15)<=$B$3,0,(-$B$2/$B$4*(D15-$B$1)))	=H14+G15
16	=IF(ABS(E16)>$B$3,"yes","no")	=IF(ABS(E16)<=$B$3,0,(-$B$2/$B$4*(D16-$B$1)))	=H15+G16
17			

	A	B	C	D	E	F	G	H	I
1	Target yt =	0							
2	lambda =	0.4							
3	L =	10							
4	g =	0.8							
5									
6	Obs	Orig_out	Orig_Nt	Adj_out_t	EWMA_t	|EWMA_t|>L?	Adj_Obs_t+1	Cum_Adj	
7	1	0	0						
8	2	16	16	16	6.400	no	0	0	Process is reset. Use original disturbance Nt to determine adjusted process output.
9	3	24	8	24	13.440	yes	-12	-12	
10	4	29	5	17	6.800	no	0	-12	
11	5	34	5	22	12.880	yes	-11	-23	
12	6	24	-10	1	0.400	no	0	-23	
13	7	31	7	8	3.440	no	0	-23	
14	8	26	-5	3	3.264	no	0	-23	
15	9	38	12	15	7.958	no	0	-23	
16	10	29	-9	6	7.175	no	0	-23	
17	...								

11-3 continued

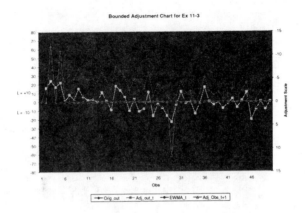

The chart with $\lambda = 0.4$ exhibits less variability, but is further from target on average than for the chart with $\lambda = 0.3$.

11-5.   The Variogram. Consider the variance of observations that are m periods apart; that is, $V_m = V(y_{t+m} - y_t)$. A graph of V_m/V_1 versus m is called a variogram. It is a nice way to check a data series for nonstationary (drifting mean) behavior. If a data series is completely uncorrelated (white noise) the variogram will always produce a plot that stays near unity. If the data series is autocorrelated but stationary, the plot of the variogram will increase for a while, but as m increases the plot of V_m/V_1 will gradually stabilize and not increase any further. The plot of V_m/V_1 versus m will increase without bound for nonstationary data. Apply this technique to the data in Table 11-1. Is there an indication of nonstationary behavior? Calculate the sample autocorrelation function for the data. Compare the interpretation of both graphs.

This exercise may be solved in Excel.

	A	B	C	D	E	F	G	H	I	J	K
1	t	Yt / m =>	1	2	3	4	5	6	7	8	9
2	1	0									
3	2	16	=B3-B2								
4	3	24	=B4-B3	=B4-B2							
5	4	29	=B5-B4	=B5-B3	=B5-B2						
6	5	34	=B6-B5	=B6-B4	=B6-B3	=B6-B2					
7	6	24	=B7-B6	=B7-B5	=B7-B4	=B7-B3	=B7-B2				
8	7	31	=B8-B7	=B8-B6	=B8-B5	=B8-B4	=B8-B3	=B8-B2			
9	8	26	=B9-B8	=B9-B7	=B9-B6	=B9-B5	=B9-B4	=B9-B3	=B9-B2		
10	9	38	=B10-B9	=B10-B8	=B10-B7	=B10-B6	=B10-B5	=B10-B4	=B10-B3	=B10-B2	
11	10	29	=B11-B10	=B11-B9	=B11-B8	=B11-B7	=B11-B6	=B11-B5	=B11-B4	=B11-B3	=B11-B2
12	...										

	A	B	C	D	E	F	G	H	I	J	K
1	t	Yt / m =>	1	2	3	4	5	6	7	8	9
2	1		0								
3	2		16	16							
4	3		24	8	24						
5	4		29	5	13	29					
6	5		34	5	10	18	34				
7	6		24	-10	-5	0	8	24			
8	7		31	7	-3	2	7	15	31		
9	8		26	-5	2	-8	-3	2	10	26	
10	9		38	12	7	14	4	9	14	22	38
11	10		29	-9	3	-2	5	-5	0	5	13	29
12	...										

11-5 continued

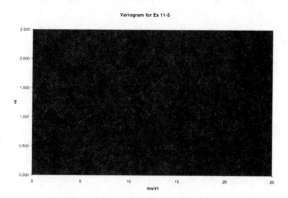

MTB > Stat > Time Series > Autocorrelation Function

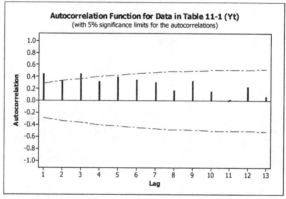

Autocorrelation Function: Yt

Lag	ACF	T	LBQ
1	0.440855	3.12	10.31
2	0.334961	2.01	16.39
3	0.440819	2.45	27.14
4	0.316478	1.58	32.80
5	0.389094	1.85	41.55
6	0.345327	1.54	48.59
7	0.299822	1.28	54.03
8	0.164698	0.68	55.71
9	0.325056	1.33	62.41
10	0.149321	0.59	63.86
11	0.012158	0.05	63.87
12	0.228540	0.90	67.44
13	0.066173	0.26	67.75

Variogram appears to be increasing, so the observations are correlated and there may be some mild indication of nonstationary behavior. The slow decline in the sample ACF also indicates the data are correlated and potentially nonstationary.

11-7. Use the data in Exercise 11-6 to construct a bounded adjustment chart. Use $\lambda = 0.2$ and set $L = 12$. How does the bounded adjustment chart perform relative to the integral control adjustment procedure in part (a) of Exercise 11-6?

Observation, t	y_t	Observation. t	y_t				
1	215.8	26	171.9	13	174.3	38	160.0
2	195.8	27	170.4	14	166.5	39	132.9
3	191.3	28	169.4	15	157.3	40	152.8
4	185.3	29	170.9	16	166.6	41	143.7
5	216.0	30	157.2	17	160.6	42	152.3
6	176.9	31	172.4	18	155.6	43	111.3
7	176.0	32	160.7	19	152.5	44	143.6
8	162.6	33	145.6	20	164.9	45	129.9
9	187.5	34	159.9	21	159.0	46	122.9
10	180.5	35	148.6	22	174.2	47	126.2
11	174.5	36	151.1	23	143.6	48	133.2
12	151.6	37	162.1	24	163.1	49	145.0
				25	189.7	50	129.5

This exercise may be solved in Excel.

	A	B	C	D	E
1	Target yt =	200			
2	lambda =	0.2			
3	L =	12			
4	g =	1.2			
5					
6	Obs, t	Orig_out	Orig_Nt	Adj_out_t	EWMA_t
7	1	215.8	0		
8	2	195.8	=B8-B7	=B8	=IF(ABS(E7)<=$B$3,$B$2*(D8-$B$1)+(1-$B$2)*E7,$B$2*(D8-$B$1)+(1-$B$2)*0)
9	3	191.3	=B9-B8	=D8+C9+G8	=IF(ABS(E8)<=$B$3,$B$2*(D9-$B$1)+(1-$B$2)*E8,$B$2*(D9-$B$1)+(1-$B$2)*0)
10	4	185.3	=B10-B9	=D9+C10+G9	=IF(ABS(E9)<=$B$3,$B$2*(D10-$B$1)+(1-$B$2)*E9,$B$2*(D10-$B$1)+(1-$B$2)*0)
11	5	216	=B11-B10	=D10+C11+G10	=IF(ABS(E10)<=$B$3,$B$2*(D11-$B$1)+(1-$B$2)*E10,$B$2*(D11-$B$1)+(1-$B$2)*0)
12	6	176.9	=B12-B11	=D11+C12+G11	=IF(ABS(E11)<=$B$3,$B$2*(D12-$B$1)+(1-$B$2)*E11,$B$2*(D12-$B$1)+(1-$B$2)*0)
13	7	176	=B13-B12	=D12+C13+G12	=IF(ABS(E12)<=$B$3,$B$2*(D13-$B$1)+(1-$B$2)*E12,$B$2*(D13-$B$1)+(1-$B$2)*0)
14	8	162.6	=B14-B13	=D13+C14+G13	=IF(ABS(E13)<=$B$3,$B$2*(D14-$B$1)+(1-$B$2)*E13,$B$2*(D14-$B$1)+(1-$B$2)*0)
15	9	187.5	=B15-B14	=D14+C15+G14	=IF(ABS(E14)<=$B$3,$B$2*(D15-$B$1)+(1-$B$2)*E14,$B$2*(D15-$B$1)+(1-$B$2)*0)
16	10	180.5	=B16-B15	=D15+C16+G15	=IF(ABS(E15)<=$B$3,$B$2*(D16-$B$1)+(1-$B$2)*E15,$B$2*(D16-$B$1)+(1-$B$2)*0)
17	...				

	F	G	H	I
1				
2				
3				
4				
5				
6	|EWMA_t| >L?	Adj_Obs_t+1	Cum_Adj	
7				
8	=IF(ABS(E8)>$B$3,"yes","no")	=IF(ABS(E8)<=$B$3,0,(-$B$2/$B$4*(D8-$B$1)))	=H7+G8	
9	=IF(ABS(E9)>$B$3,"yes","no")	=IF(ABS(E9)<=$B$3,0,(-$B$2/$B$4*(D9-$B$1)))	=H8+G9	
10	=IF(ABS(E10)>$B$3,"yes","no")	=IF(ABS(E10)<=$B$3,0,(-$B$2/$B$4*(D10-$B$1)))	=H9+G10	
11	=IF(ABS(E11)>$B$3,"yes","no")	=IF(ABS(E11)<=$B$3,0,(-$B$2/$B$4*(D11-$B$1)))	=H10+G11	
12	=IF(ABS(E12)>$B$3,"yes","no")	=IF(ABS(E12)<=$B$3,0,(-$B$2/$B$4*(D12-$B$1)))	=H11+G12	
13	=IF(ABS(E13)>$B$3,"yes","no")	=IF(ABS(E13)<=$B$3,0,(-$B$2/$B$4*(D13-$B$1)))	=H12+G13	
14	=IF(ABS(E14)>$B$3,"yes","no")	=IF(ABS(E14)<=$B$3,0,(-$B$2/$B$4*(D14-$B$1)))	=H13+G14	Process is reset. Use original disturbance Nt to determine adjusted process output.
15	=IF(ABS(E15)>$B$3,"yes","no")	=IF(ABS(E15)<=$B$3,0,(-$B$2/$B$4*(D15-$B$1)))	=H14+G15	
16	=IF(ABS(E16)>$B$3,"yes","no")	=IF(ABS(E16)<=$B$3,0,(-$B$2/$B$4*(D16-$B$1)))	=H15+G16	
17				

11-7 continued

	A	B	C	D	E	F	G	H	I
1	Target yt =	200							
2	lambda =	0.2							
3	L =	12							
4	g =	1.2							
5									
6	Obs, t	Orig_out	Orig_Nt	Adj_out _t	EWMA _t	|EWMA_t| >L?	Adj_Obs _t+1	Cum_Adj	
7	1	215.8	0						
8	2	195.8	-20	196	-0.840	no	0.0	0.0	
9	3	191.3	-4.5	191.300	-2.412	no	0.0	0.000	
10	4	185.3	-6	185.300	-4.870	no	0.000	0.000	
11	5	216.0	30.7	216.000	-0.696	no	0.000	0.000	
12	6	176.9	-39.1	176.900	-5.177	no	0.000	0.000	
13	7	176.0	-0.9	176.000	-8.941	no	0.000	0.000	
14	8	162.6	-13.4	162.600	-14.633	yes	6.233	6.233	Process is reset. Use original disturbance Nt to determine adjusted process output.
15	9	187.5	24.9	193.733	-1.253	no	0.000	6.233	
16	10	180.5	-7	186.733	-3.656	no	0.000	6.233	
17	...								

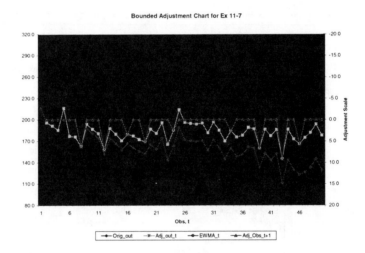

Bounded Adjustment Chart for Ex 11-7

Behavior of the bounded adjustment control scheme is similar to both integral control schemes ($\lambda = 0.2$ and $\lambda = 0.4$).

11-9.   Consider the observations in the following table. The target value for this process is 50.

Observation, t	y_t	Observation, t	y_t				
1	50	26	43	13	67	38	27
2	58	27	39	14	55	39	29
3	54	28	32	15	56	40	35
4	45	29	37	16	65	41	27
5	56	30	44	17	65	42	33
6	56	31	52	18	61	43	25
7	66	32	42	19	57	44	21
8	55	33	47	20	61	45	16
9	69	34	33	21	64	46	24
10	56	35	49	22	43	47	18
11	63	36	34	23	44	48	20
12	54	37	40	24	45	49	23
				25	39	50	26

(a) Set up an integral controller for this process. Assume that the gain for the adjustment variable is g = 1.6 and assume that λ = 0.2 in the EWMA forecasting procedure will provide adequate one-step-ahead predictions.
(b) How much reduction in variability around the target does the integral controller achieve?

This exercise may be solved in Excel.

	A	B	C	D	E	F	G
1	T =	50					
2	lambda =	0.2					
3	g =	1.6					
4							
5	Obs	Orig_out	Orig_Nt	Adj_out_t	Adj_Obs_t+1	Cum_Adj	
6	1	50					
7	2	58	=B7-B6	=B7	=-$B$2/$B$3*(D7-$B$1)	=F6+E7	Process is reset. Use original disturbance Nt to determine adjusted process output.
8	3	54	=B8-B7	=D7+C8+E7	=-$B$2/$B$3*(D8-$B$1)	=F7+E8	
9	4	45	=B9-B8	=D8+C9+E8	=-$B$2/$B$3*(D9-$B$1)	=F8+E9	
10	5	56	=B10-B9	=D9+C10+E9	=-$B$2/$B$3*(D10-$B$1)	=F9+E10	
11	6	56	=B11-B10	=D10+C11+E10	=-$B$2/$B$3*(D11-$B$1)	=F10+E11	
12	7	66	=B12-B11	=D11+C12+E11	=-$B$2/$B$3*(D12-$B$1)	=F11+E12	
13	8	55	=B13-B12	=D12+C13+E12	=-$B$2/$B$3*(D13-$B$1)	=F12+E13	
14	9	69	=B14-B13	=D13+C14+E13	=-$B$2/$B$3*(D14-$B$1)	=F13+E14	
15	10	56	=B15-B14	=D14+C15+E14	=-$B$2/$B$3*(D15-$B$1)	=F14+E15	
16	...						
55	49	23	=B55-B54	=D54+C55+E54	=-$B$2/$B$3*(D55-$B$1)	=F54+E55	
56	50	26	=B56-B55	=D55+C56+E55	=-$B$2/$B$3*(D56-$B$1)	=F55+E56	
57							
58		Unadjusted		Adjusted			
59	SS =	=SUMSQ(B6:B56)		=SUMSQ(D6:D56)			
60	Average =	=AVERAGE(B6:B56)		=AVERAGE(D6:D56)			
61	Variance =	=VAR(B6:B56)		=VAR(D6:D56)			

11-9 continued

	A	B	C	D	E	F	G
1	T =	50					
2	lambda =	0.2					
3	g =	1.6					
4							
5	Obs	Orig_out	Orig_Nt	Adj_out_t	Adj_Obs_t+1	Cum_Adj	
6	1	50					
7	2	58	8.0	58.0	-1.0	-1.0	Process is reset. Use original disturbance Nt to determine adjusted process output.
8	3	54	-4.0	53.0	-0.4	-1.4	
9	4	45	-9.0	43.6	0.8	-0.6	
10	5	56	11.0	55.4	-0.7	-1.3	
11	6	56	0.0	54.7	-0.6	-1.8	
12	7	66	10.0	64.2	-1.8	-3.6	
13	8	55	-11.0	51.4	-0.2	-3.8	
14	9	69	14.0	65.2	-1.9	-5.7	
15	10	56	-13.0	50.3	0.0	-5.7	
16	...						
55	49	23	3.0	45.1	0.6	22.7	
56	50	26	3.0	48.7	0.2	22.9	
57							
58		Unadjusted		Adjusted			
59	SS =	109,520		108,629			
60	Average =	44.4		46.262			
61	Variance =	223.51		78.32			

Based on the variance, there is a significant reduction in variability with use of an integral control scheme.

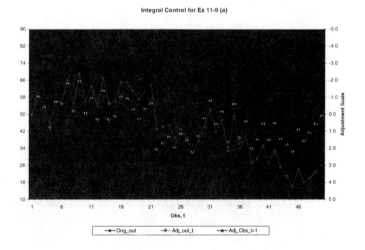

Integral Control for Ex 11-9 (a)

11-9 continued

(c) Rework parts (a) and (b) assuming that $\lambda = 0.4$. What change does this make in the variability around the target in comparison to that achieved with $\lambda = 0.2$?

	A	B	C	D	E	F	G
1	T =	50					
2	lambda =	0.4					
3	g ±	1.6					
4							
5	Obs	Orig_out	Orig_Nt	Adj_out_t	Adj_Obs_t+1	Cum_Adj	
6	1	50					
7	2	58	8.0	58.0	-2.0	-2.0	Process is reset. Use original disturbance Nt to determine adjusted process output.
8	3	54	-4.0	52.0	-0.5	-2.5	
9	4	45	-9.0	42.5	1.9	-0.6	
10	5	56	11.0	55.4	-1.3	-2.0	
11	6	56	0.0	54.0	-1.0	-3.0	
12	7	66	10.0	63.0	-3.3	-6.2	
13	8	55	-11.0	48.8	0.3	-5.9	
14	9	69	14.0	63.1	-3.3	-9.2	
15	10	56	-13.0	46.8	0.8	-8.4	
16	...						
51	45	16	-5.0	37.4	3.1	24.6	
52	46	24	8.0	48.6	0.4	24.9	
53	47	18	-6.0	42.9	1.8	26.7	
54	48	20	2.0	46.7	0.8	27.5	
55	49	23	3.0	50.5	-0.1	27.4	
56	50	26	3.0	53.4	-0.8	26.5	
57							
58	SS =	109,520		114,819			
59	Average =	44.4		47.833			
60	Variance =	223.51		56.40			

There is a slight reduction in variability with use of $\lambda = 0.4$, as compared to $\lambda = 0.2$, with a process average slightly closer to the target of 50.

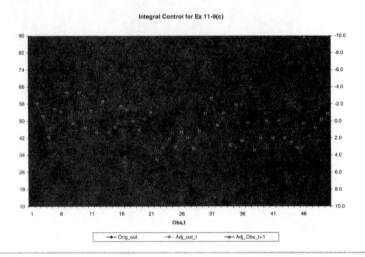

CHAPTER 12

Factorial and Fractional Factorial Experiments for Process Design and Improvement

Learning Objectives

After completing this chapter you should be able to:

1. Explain how designed experiments can be used to improve product design and improve process performance
2. Explain how designed experiments can be used to reduce the cycle time required to develop new products and processes
3. Understand how main effects and interactions of factors can be estimated
4. Understand the factorial design concept
5. Know how to use the analysis of variance (ANOVA) to analyze data from factorial designs
6. Know how residuals are used for model adequacy checking for factorial designs
7. Know how to use the 2^k system of factorial designs
8. Know how to construct and interpret contour plots and response surface plots
9. Know how to add center points to a 2^k factorial design to test for curvature and provide an estimate of pure experimental error
10. Understand how the blocking principal can be used in a factorial design to eliminate the effects of a nuisance factor
11. Know hw to use the 2^{k-p} system of fractional factorial designs

Important Terms and Concepts

2^k factorial designs

Aliasing

Analysis procedure for factorial designs

Center points in a 2^k and 2^{k-p} factorial designs

Confounding

Controllable process variables

Defining relation for a fractional factorial design

Fractional factorial design

Guidelines for planning experiments

Main effect of a factor

Orthogonal design

Projection of 2^k and 2^{k-p} factorial designs

Residual analysis

Resolution of a fractional factorial design

Screening Experiments

Sparsity of effects principle

2^{k-p} fractional factorial designs

Analysis of variance (ANOVA)

Blocking

Completely randomized design

Contour plot

Curvature in the response function

Factorial design

Generators for a fractional factorial design

Interaction

Normal probability plot of effects

Pre-experimental planning

Regression model representation of experimental results

Residuals

Response surface

Sequential experimentation

Two-factor interaction

Exercises

Note: To analyze an experiment in MINITAB, the initial experimental layout must be created in MINITAB or defined by the user. The Excel data sets contain only the data given in the textbook; therefore some information required by MINITAB is not included. Detailed MINITAB instructions are provided for Exercises 12-1 and 12-2 to define and create designs. The remaining exercises are worked in a similar manner, and only the solutions are provided.

12-1.

An article in Industrial Quality Control (1956, pp. 5–8) describes an experiment to investigate the effect of glass type and phosphor type on the brightness of a television tube. The response measured is the current necessary (in microamps) to obtain a specified brightness level. The data are shown here. Analyze the data and draw conclusions.

	Phosphor Type		
Glass Type	1	2	3
1	280	300	290
	290	310	285
	285	295	290
2	230	260	220
	235	240	225
	240	235	230

This experiment is three replicates of a factorial design in two factors—two levels of glass type and three levels of phosphor type—to investigate brightness. Enter the data into the MINITAB worksheet using the first three columns: one column for glass type, one column for phosphor type, and one column for brightness. This is how the Excel file is structured (**Chap12.xls**). Since the experiment layout was not created in MINITAB, the design must be defined before the results can be analyzed.

After entering the data in MINITAB, select **Stat>DOE>Factorial>Define Custom Factorial Design**. Select the two factors (Glass Type and Phosphor Type), then for this exercise, check "**General full factorial**". The dialog box should look:

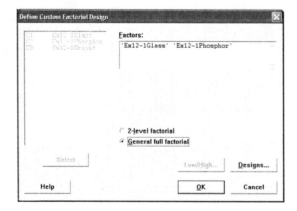

12-1 continued

Next, select "**Designs**". For this exercise, no information is provided on standard order, run order, point type, or blocks, so leave the selections as below, and click "**OK**" twice.

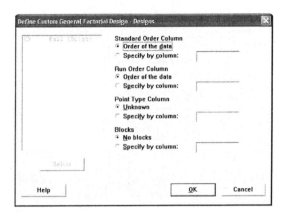

Note that MINITAB added four new columns (4 through 7) to the worksheet. DO NOT insert or delete columns between columns 1 through 7. MINITAB recognizes these contiguous seven columns as a designed experiment; inserting or deleting columns will cause the design layout to become corrupt.

Select **Stat>DOE>Factorial>Analyze Factorial Design**. Select the response (Brightness), then click on "**Terms**", verify that the selected terms are Glass Type, Phosphor Type, and their interaction, click "**OK**". Click on "**Graphs**", select "**Residuals Plots : Four in one**". The option to plot residuals versus variables is for continuous factor levels; since the factor levels in this experiment are categorical, do not select this option. Click "**OK**". Click on "**Storage**", select "**Fits**" and "**Residuals**", and click "**OK**" twice.

General Linear Model: Ex12-1Bright versus Ex12-1Glass, Ex12-1Phosphor

```
Factor            Type    Levels  Values
Ex12-1Glass       fixed      2    1, 2
Ex12-1Phosphor    fixed      3    1, 2, 3

Analysis of Variance for Ex12-1Bright, using Adjusted SS for Tests
Source                        DF    Seq SS    Adj SS    Adj MS       F      P
Ex12-1Glass                    1   14450.0   14450.0   14450.0  273.79  0.000
Ex12-1Phosphor                 2     933.3     933.3     466.7    8.84  0.004
Ex12-1Glass*Ex12-1Phosphor     2     133.3     133.3      66.7    1.26  0.318
Error                         12     633.3     633.3      52.8
Total                         17   16150.0
```

S = 7.26483   R-Sq = 96.08%   R-Sq(adj) = 94.44%

No indication of significant interaction (*P*-value is greater than 0.10). Glass type (A) and phosphor type (B) significantly affect television tube brightness (*P*-values are less than 0.10).

12-1 continued

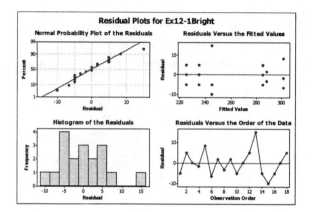

Visual examination of residuals on the normal probability plot, histogram, and versus fitted values does reveal any problems. The plot of residuals versus observation order is not meaningful since no order was provided with the data. If the model were re-fit with only Glass Type and Phosphor Type, the residuals should be re-examined.

To plot residuals versus the two factors, select **Graph>Individual Value Plot>One Y with Groups**. Select the column with stored residuals (**RESI1**) as the **Graph variable** and select one of the factors (Glass Type or Phosphor Type) as the **Categorical variable for grouping**. Click on "**Scale**", select the "**Reference Lines**" tab, and enter "**0**" for the Y axis, then click "**OK**" twice.

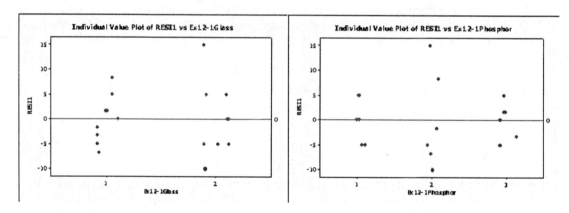

12-1 continued

Note that the plot points are "jittered" about the factor levels. To remove the jitter, select the graph to make it active then: **Editor>Select Item>Individual Symbols** and then **Editor>Edit Individual Symbols>Jitter** and de-select **Add jitter to direction**.

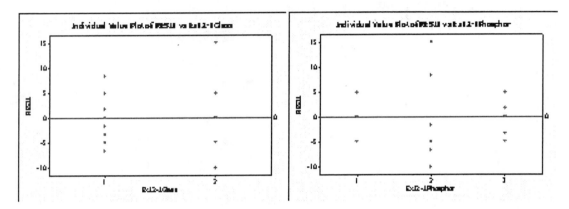

Variability appears to be the same for both glass types; however, there appears to be more variability in results with phosphor type 2.

Select **Stat>DOE>Factorial>Factorial Plots**. Select "**Interaction Plot**" and click on "**Setup**", select the response (Brightness) and both factors (Glass Type and Phosphor Type), and click "**OK**" twice.

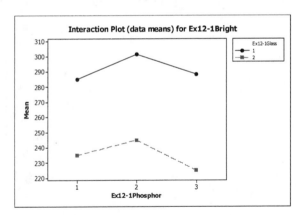

The absence of a significant interaction is evident in the parallelism of the two lines. Final selected combination of glass type and phosphor type depends on the desired brightness level.

12-1 continued

Alternate Solution: This exercise may also be solved using MINITAB's ANOVA functionality instead of its DOE functionality. The DOE functionality was selected to illustrate the approach that will be used for most of the remaining exercises. To obtain results which match the output in the textbook's Table 12.5, select **Stat>ANOVA>Two-Way**, and complete the dialog box as below.

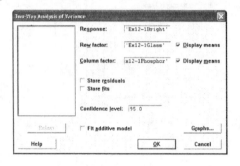

```
Two-way ANOVA: Ex12-1Bright versus Ex12-1Glass, Ex12-1Phosphor
Source            DF       SS       MS       F       P
Ex12-1Glass        1  14450.0  14450.0  273.79  0.000
Ex12-1Phosphor     2    933.3    466.7    8.84  0.004
Interaction        2    133.3     66.7    1.26  0.318
Error             12    633.3     52.8
Total             17  16150.0
S = 7.265   R-Sq = 96.08%   R-Sq(adj) = 94.44%


                        Individual 95% CIs For Mean Based on
                        Pooled StDev
Ex12-1Glass     Mean   -----+---------+---------+---------+----
1             291.667                               (--*-)
2             235.000  (--*-)
                       -----+---------+---------+---------+----
                        240       260       280       300


                        Individual 95% CIs For Mean Based on
                        Pooled StDev
Ex12-1Phosphor   Mean  -------+---------+---------+---------+--
1             260.000    (-------*-------)
2             273.333                  (-------*-------)
3             256.667  (-------*-------)
                       -------+---------+---------+---------+--
                        256.0     264.0     272.0     280.0
```

12-3.   Find the residuals from the tool life experiment in Exercise 12-2. Construct a normal probability plot of the residuals. Plot the residuals versus the predicted values. Comment on the plots.

> To find the residuals, select **Stat > DOE > Factorial > Analyze Factorial Design**. Select "**Terms**" and verify that all terms for the reduced model (A, B, C, AC) are included. Select "**Graphs**", and for residuals plots choose "**Normal plot**" and "**Residuals versus fits**". To save residuals to the worksheet, select "**Storage**" and choose "**Residuals**".

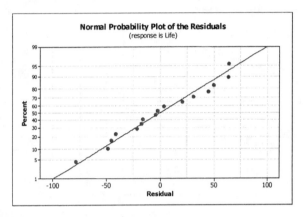

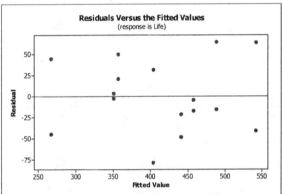

Normal probability plot of residuals indicates that the normality assumption is reasonable. Residuals versus fitted values plot shows that the equal variance assumption across the prediction range is reasonable.

12-5.   Consider the experiment in Exercise 12-4. Plot the residuals against the levels of factors A, B, C, and D. Also construct a normal probability plot of the residuals. Comment on these plots.

> To find the residuals, select **Stat > DOE > Factorial > Analyze Factorial Design**. Select "**Terms**" and verify that all terms for the reduced model are included. Select "**Graphs**", choose "**Normal plot**" of residuals and "**Residuals versus variables**", and then select the variables.

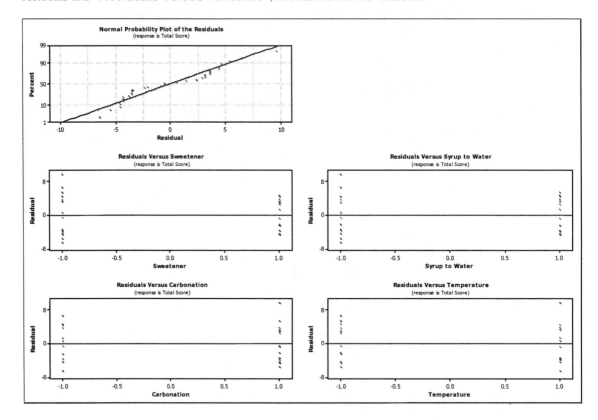

> There appears to be a slight indication of inequality of variance for sweetener and syrup ratio, as well as a slight indication of an outlier. This is not serious enough to warrant concern.

12-7. Suppose that only the data from replicate I in Exercise 12-4 were available. Analyze the data and draw appropriate conclusions.

> Create a 2^4 factorial design in MINITAB, and then enter the data. Select **Stat > DOE > Factorial > Analyze Factorial Design**. Since there is only one replicate of the experiment, select "**Terms**" and verify that all terms are selected. Then select "**Graphs**", choose the normal effects plot, and set alpha to 0.10

Factorial Fit: Total Score versus Sweetener, Syrup to Water, ...

Estimated Effects and Coefficients for Total Score (coded units)

Term	Effect	Coef
Constant		183.625
Sweetener	-10.500	-5.250
Syrup to Water	-0.250	-0.125
Carbonation	0.750	0.375
Temperature	5.500	2.750
Sweetener*Syrup to Water	4.000	2.000
Sweetener*Carbonation	1.000	0.500
Sweetener*Temperature	-6.250	-3.125
Syrup to Water*Carbonation	-1.750	-0.875
Syrup to Water*Temperature	-3.000	-1.500
Carbonation*Temperature	1.000	0.500
Sweetener*Syrup to Water*Carbonation	-7.500	-3.750
Sweetener*Syrup to Water*Temperature	4.250	2.125
Sweetener*Carbonation*Temperature	0.250	0.125
Syrup to Water*Carbonation* Temperature	-2.500	-1.250
Sweetener*Syrup to Water* Carbonation*Temperature	3.750	1.875

...

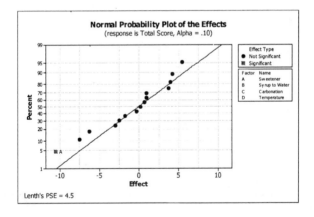

From visual examination of the normal probability plot of effects, only factor A (sweetener) is significant at $\alpha = 0.10$.

12-7 continued

Re-fit and analyze the reduced model.

```
Factorial Fit: Total Score versus Sweetener
Estimated Effects and Coefficients for Total Score (coded units)
Term           Effect      Coef   SE Coef      T      P
Constant               183.625     1.865   98.48  0.000
Sweetener   -10.500    -5.250     1.865   -2.82  0.014

S = 7.45822    R-Sq = 36.15%    R-Sq(adj) = 31.59%

Analysis of Variance for Total Score (coded units)
Source            DF    Seq SS    Adj SS   Adj MS      F      P
Main Effects       1    441.00   441.000   441.00   7.93  0.014
Residual Error    14    778.75   778.750    55.63
  Pure Error      14    778.75   778.750    55.63
Total             15   1219.75
```

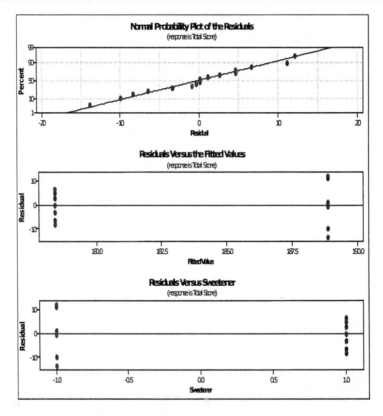

There appears to be a slight indication of inequality of variance for sweetener, as well as in the predicted values. This is not serious enough to warrant concern.

A reduced model of Factor A (type of sweetener) is sufficient to model taste.

12-9.   Show how a 2^5 experiment could be set up in two blocks of 16 runs each. Specifically, which runs would be made in each block?

A 2^5 design in two blocks will lose the ABCDE interaction to blocks.

Block 1		Block 2	
(1)	ae	a	e
ab	be	b	abe
ac	ce	c	ace
bc	abce	abc	bce
ad	de	d	ade
bd	abde	abd	bde
cd	acde	acd	cde
abcd	bcde	bcd	abcde

12-11. An article in Industrial and Engineering Chemistry ("More on Planning Experiments to Increase Research Efficiency," 1970, pp. 60–65) uses a 2^{5-2} design to investigate the effect of A = condensation temperature, B = amount of material 1, C = solvent volume, D = condensation time, and E = amount of material 2, on yield. The results obtained are as follows:

$$
\begin{aligned}
e &= 23.2 & cd &= 23.8 \\
ab &= 15.5 & ace &= 23.4 \\
ad &= 16.9 & bde &= 16.8 \\
bc &= 16.2 & abcde &= 18.1
\end{aligned}
$$

(a) Verify that the design generators used were $I = ACE$ and $I = BDE$.
(b) Write down the complete defining relation and the aliases from this design.

Enter the factor levels and yield data into a MINITAB worksheet, then define the experiment using **Stat>DOE>Factorial>Define Custom Factorial Design**.

Select **Stat>DOE>Factorial>Analyze Factorial Design**. Since there is only one replicate of the experiment, select "**Terms**" and verify that all main effects and two-factor interaction effects are selected.

```
Factorial Fit: yield versus A:Temp, B:Matl1, C:Vol, D:Time, E:Matl2
Estimated Effects and Coefficients for yield (coded units)
Term             Effect    Coef
Constant                   19.238
A:Temp           -1.525    -0.762
B:Matl1          -5.175    -2.587
C:Vol             2.275     1.138
D:Time           -0.675    -0.337
E:Matl2           2.275     1.138
A:Temp*B:Matl1    1.825     0.913
A:Temp*D:Time    -1.275    -0.638
...
Alias Structure
I + A:Temp*C:Vol*E:Matl2 + B:Matl1*D:Time*E:Matl2 + A:Temp*B:Matl1*C:Vol*D:Time
A:Temp + C:Vol*E:Matl2 + B:Matl1*C:Vol*D:Time + A:Temp*B:Matl1*D:Time*E:Matl2
B:Matl1 + D:Time*E:Matl2 + A:Temp*C:Vol*D:Time + A:Temp*B:Matl1*C:Vol*E:Matl2
C:Vol + A:Temp*E:Matl2 + A:Temp*B:Matl1*D:Time + B:Matl1*C:Vol*D:Time*E:Matl2
D:Time + B:Matl1*E:Matl2 + A:Temp*B:Matl1*C:Vol + A:Temp*C:Vol*D:Time*E:Matl2
E:Matl2 + A:Temp*C:Vol + B:Matl1*D:Time + A:Temp*B:Matl1*C:Vol*D:Time*E:Matl2
A:Temp*B:Matl1 + C:Vol*D:Time + A:Temp*D:Time*E:Matl2 + B:Matl1*C:Vol*E:Matl2
A:Temp*D:Time + B:Matl1*C:Vol + A:Temp*B:Matl1*E:Matl2 + C:Vol*D:Time*E:Matl2
```

From the Alias Structure shown in the Session Window, the complete defining relation is:
$I = ACE = BDE = ABCD$.

The aliases are:
$$A*I = A*ACE = A*BDE = A*ABCD \Rightarrow A = CE = ABDE = BCD$$
$$B*I = B*ACE = B*BDE = B*ABCD \Rightarrow B = ABCE = DE = ACD$$
$$C*I = C*ACE = C*BDE = C*ABCD \Rightarrow C = AE = BCDE = ABD$$
$$\cdots$$
$$*I = AB*ACE = AB*BDE = AB*ABCD \Rightarrow AB = BCE = ADE = CD$$

The remaining aliases are calculated in a similar fashion.

12-11 continued

(c) Estimate the main effects.

A	B	C	D	E	yield
-1	-1	-1	-1	1	23.2
1	1	-1	-1	-1	15.5
1	-1	-1	1	-1	16.9
-1	1	1	-1	-1	16.2
-1	-1	1	1	-1	23.8
1	-1	1	-1	1	23.4
-1	1	-1	1	1	16.8
1	1	1	1	1	18.1

$[A] = A + CE + BCD + ABDE$
$\qquad = \frac{1}{4}(-23.2 + 15.5 + 16.9 - 16.2 - 23.8 + 23.4 - 16.8 + 18.1) = \frac{1}{4}(-6.1) = -1.525$
$[AB] = AB + BCE + ADE + CD$
$\qquad = \frac{1}{4}(+23.2 + 15.5 - 16.9 - 16.2 + 23.8 - 23.4 - 16.8 + 18.1) = \frac{1}{4}(7.3) = 1.825$

This are the same effect estimates provided in the MINITAB output above. The other main effects and interaction effects are calculated in the same way.

(d) Prepare an analysis of variance table. Verify that the AB and AD interactions are available to use as error. Select **Stat > DOE > Factorial > Analyze Factorial Design**. Since there is only a single replicate of the experiment, select "**Terms**" and verify that all main effects and two-factor interaction effects are selected. Then select "**Graphs**", choose the normal effects plot, and set alpha to 0.10.

```
Factorial Fit: yield versus A:Temp, B:Matl1, C:Vol, D:Time, E:Matl2
...
Analysis of Variance for yield (coded units)
Source              DF  Seq SS  Adj SS  Adj MS   F   P
Main Effects         5  79.826  79.826  15.965   *   *
2-Way Interactions   2   9.913   9.913   4.956   *   *
Residual Error       0       *       *       *
Total                7  89.739
...
```

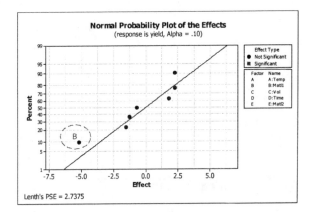

Although none of the effects is significant at 0.10, main effect B (amount of material 1) is more than twice as large as the 2nd largest effect (absolute values) and falls far from a line passing through the remaining points. Re-fit a reduced model containing only the B main effect, and pool the remaining terms to estimate error.

12-11 continued

Select **Stat > DOE > Factorial > Analyze Factorial Design**. Select "**Terms**" and select "**B**". Then select "**Graphs**", and select the "**Normal plot**" and "**Residuals versus fits**" residual plots.

```
Factorial Fit: yield versus B:Matl1
Estimated Effects and Coefficients for yield (coded units)
Term        Effect    Coef   SE Coef      T       P
Constant             19.238   0.8682   22.16   0.000
B:Matl1    -5.175   -2.587   0.8682   -2.98   0.025
...
Analysis of Variance for yield (coded units)
Source           DF   Seq SS   Adj SS   Adj MS     F       P
Main Effects      1    53.56    53.56   53.561   8.88   0.025
Residual Error    6    36.18    36.18    6.030
  Pure Error      6    36.18    36.18    6.030
Total             7    89.74
...
```

(e) Plot the residuals versus the fitted values. Also construct a normal probability plot of the residuals. Comment on the results.

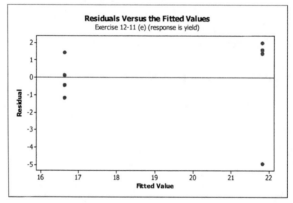

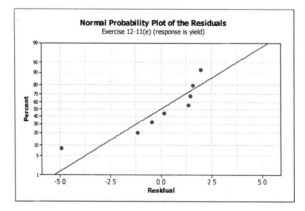

Residual plots indicate a potential outlier. The run should be investigated for any issues which occurred while running the experiment. If no issues can be identified, it may be necessary to make additional experimental runs

12-13. Reconsider the data in Exercise 12-12. Suppose that four center points were added to this experiment. The molecular weights at the center point are 90, 87, 86, and 93.

$$
\begin{array}{ll}
(1) = 88 & d = 86 \\
a = 80 & ad = 81 \\
b = 89 & bd = 85 \\
ab = 87 & abd = 86 \\
c = 86 & cd = 85 \\
ac = 81 & acd = 79 \\
bc = 82 & bcd = 84 \\
abc = 80 & abcd = 81
\end{array}
$$

(a) Analyze the data as you did in Exercise 12-12, but include a test for curvature.

Create a 2^4 factorial design with four center points in MINITAB, and then enter the data. Select **Stat > DOE > Factorial > Analyze Factorial Design**. Select "**Terms**" and verify that all main effects and two-factor interactions are selected. Also, DO NOT include the center points in the model (uncheck the default selection). This will ensure that if both lack of fit and curvature are not significant, the main and interaction effects are tested for significance against the correct residual error (lack of fit + curvature + pure error). See the dialog box below.

To summarize MINITAB's functionality, curvature is always tested against pure error and lack of fit (if available), regardless of whether center points are included in the model. The inclusion/exclusion of center points in the model affects the total residual error used to test significance of effects. Assuming that lack of fit and curvature tests are not significant, all three (curvature, lack of fit, and pure error) should be included in the residual mean square.

When looking at results in the ANOVA table, the first test to consider is the "lack of fit" test, which is a test of significance for terms not included in the model (in this exercise, the three-factor and four-factor interactions). If lack of fit is significant, the model is not correctly specified, and some terms need to be added to the model.

If lack of fit is not significant, the next test to consider is the "curvature" test, which is a test of significance for the pure quadratic terms. If this test is significant, no further statistical analysis should be performed because the model is inadequate.

12-13 continued

If tests for both lack of fit and curvature are not significant, then it is reasonable to pool the curvature, pure error, and lack of fit (if available) and use this as the basis for testing for significant effects. (In MINITAB, this is accomplished by not including center points in the model.)

```
Factorial Fit: Mole Wt versus A, B, C, D
Estimated Effects and Coefficients for Mole Wt (coded units)
Term       Effect    Coef   SE Coef     T       P
Constant           848.00    8.521    99.52   0.000
A          -37.50  -18.75    9.527    -1.97   0.081
B           10.00    5.00    9.527     0.52   0.612
C          -30.00  -15.00    9.527    -1.57   0.150
D           -7.50   -3.75    9.527    -0.39   0.703
A*B         22.50   11.25    9.527     1.18   0.268
A*C         -2.50   -1.25    9.527    -0.13   0.898
A*D          5.00    2.50    9.527     0.26   0.799
B*C        -20.00  -10.00    9.527    -1.05   0.321
B*D          2.50    1.25    9.527     0.13   0.898
C*D          7.50    3.75    9.527     0.39   0.703
...

Analysis of Variance for Mole Wt (coded units)
Source              DF  Seq SS  Adj SS  Adj MS      F       P
Main Effects         4    9850    9850  2462.5    1.70   0.234
2-Way Interactions   6    4000    4000   666.7    0.46   0.822
Residual Error       9   13070   13070  1452.2
  Curvature          1    8820    8820  8820.0   16.60   0.004 ***
  Lack of Fit        5    1250    1250   250.0    0.25   0.915
  Pure Error         3    3000    3000  1000.0
Total               19   26920
...
```

(b) If curvature is significant in an experiment such as this one, describe what strategy you would pursue next to improve your model of the process.

The test for curvature is significant (*P*-value = 0.004). Although one could pick a "winning combination" from the experimental runs, a better strategy is to add runs that would enable estimation of the quadratic effects. This approach to sequential experimentation is presented in Chapter 13.

CHAPTER 13

Process Optimization with Designed Experiments

Learning Objectives

After completing this chapter you should be able to:
1. Explain how designed experiments can be used to improve product design and improve process performance
2. Explain how designed experiments can be used to reduce the cycle time required to develop new products and processes
3. Understand how main effects and interactions of factors can be estimated
4. Understand the factorial design concept
5. Know how to use the analysis of variance (ANOVA) to analyze data from factorial designs
6. Know how residuals are used for model adequacy checking for factorial designs
7. Know how to use the 2^k system of factorial designs
8. Know how to construct and interpret contour plots and response surface plots
9. Know how to add center points to a 2^k factorial design to test for curvature and provide an estimate of pure experimental error
10. Understand how the blocking principal can be used in a factorial design to eliminate the effects of a nuisance factor
11. Know hw to use the 2^{k-p} system of fractional factorial designs

Important Terms and Concepts

Central composite design

Contour plot

Crossed array design

EVOP cycle

First-order model

Method of steepest ascent

Outer array design

Process robustness study

Response surface

Robust parameter design (RPD)

Second-order model

Taylor series

Combined array design

Controllable variable

Evolutionary operation (EVOP

EVOP phase

Inner array design

Noise variable

Path of steepest ascent

Response model

Response surface methodology (RSM)

Rotatable design

Sequential experimentation

Transmission of error

Exercises

Note: To analyze an experiment in MINITAB, the initial experimental layout must be created in MINITAB or defined by the user. The Excel data sets contain only the data given in the textbook; therefore some information required by MINITAB is not included. The MINITAB instructions provided for the factorial designs in Chapter 12 are similar to those for response surface designs in this Chapter.

13-1. Consider the first-order model $\hat{y}_t = 75 + 10x_1 + 6x_2$

(a) Sketch the contours of constant predicted response over the range $-1 \le x_i \le +1$, $i = 1, 2$.

In MINITAB, designate columns for x_1 and x_2, generate a grid of values from -1 to $+1$, then calculation y.

Graph>Contour Plot

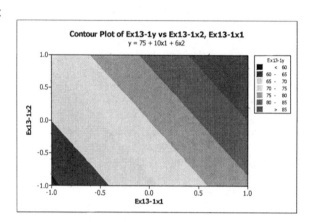

(b) Find the direction of steepest ascent.

$$\hat{y} = 75 + 10x_1 + 6x_2 \quad -1 \le x_1 \le 1; 1 \le x_2 \le 1$$

$$\frac{x_2}{x_1} = \frac{6}{10} = 0.6$$

$$\Delta x_1 = 1$$

$$\Delta x_2 = 0.6$$

13-3.   An experiment was run to study the effect of two factors, time and temperature, on the inorganic impurity levels in paper pulp. The results of this experiment are shown here:

x_1	x_2	y
-1	-1	210
1	-1	95
-1	1	218
1	1	100
-1.5	0	225
1.5	0	50
0	-1.5	175
0	1.5	180
0	0	145
0	0	175
0	0	158
0	0	166

(a) What type of experimental design has been used in this study? Is the design rotatable?

This design is a CCD with $k = 2$ and $\alpha = 1.5$. The design is not rotatable.

(b) Fit a quadratic model to the response, using the method of least squares.

Enter the factor levels and response data into a MINITAB worksheet, including a column indicating whether a run is a center point run (1 = not center point, 0 = center point). Then define the experiment using **Stat>DOE>Response Surface>Define Custom Response Surface Design**. Select **Stat>DOE>Response Surface>Analyze Response Surface Design**. Select "**Terms**" and verify that all main effects, two-factor interactions, and quadratic terms are selected.

```
Response Surface Regression: y versus x1, x2
The analysis was done using coded units.
Estimated Regression Coefficients for y
Term          Coef  SE Coef        T      P
Constant   160.868    4.555   35.314  0.000
x1         -87.441    4.704  -18.590  0.000
x2           3.618    4.704    0.769  0.471
x1*x1      -24.423    7.461   -3.273  0.017
x2*x2       15.577    7.461    2.088  0.082
x1*x2       -1.688   10.285   -0.164  0.875
...
Analysis of Variance for y
Source           DF    Seq SS    Adj SS    Adj MS       F      P
Regression        5   30583.4   30583.4    6116.7   73.18  0.000
  Linear          2   28934.2   28934.2   14467.1  173.09  0.000
  Square          2    1647.0    1647.0     823.5    9.85  0.013
  Interaction     1       2.3       2.3       2.3    0.03  0.875
Residual Error    6     501.5     501.5      83.6
  Lack-of-Fit     3      15.5      15.5       5.2    0.03  0.991
  Pure Error      3     486.0     486.0     162.0
Total            11   31084.9
...
Estimated Regression Coefficients for y using data in uncoded units
Term          Coef
Constant   160.8682
x1         -58.2941
x2           2.4118
x1*x1      -10.8546
x2*x2        6.9231
x1*x2       -0.7500
```

The model is $y = 160.9 - 58.3x_1 + 2.4x_2 - 10.9x_1^2 + 6x_2^2 - 0.75x_1x_2$

13-3 continued

(c) Construct the fitted impurity response surface. What values of x_1 and x_2 would you recommend if you wanted to minimize the impurity level?

Stat>DOE>Response Surface>Contour/Surface Plots

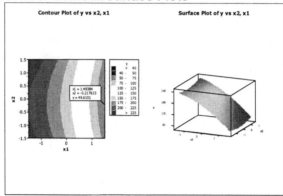

From visual examination of the contour and surface plots, it appears that minimum purity can be achieved by setting x_1 (time) = +1.5 and letting x_2 (temperature) range from −1.5 to + 1.5. The range for x_2 agrees with the ANOVA results indicating that it is statistically insignificant (P-value = 0.471). The level for temperature could be established based on other considerations, such as cost. A flag is planted at one option on the contour plot above.

(d) Suppose that

$$x_1 = \frac{temp - 750}{50} \qquad x_2 = \frac{time - 30}{15}$$

where temperature is in °C and time is in hours. Find the optimum operating conditions in terms of the natural variables temperature and time.

$$\text{Temp} = 50x_1 + 750 = 50(+1.50) + 750 = 825$$

$$\text{Time} = 15x_2 + 30 = 15(-0.22) + 30 = 26.7$$

13-5.   An article in Rubber Chemistry and Technology (Vol. 47, 1974, pp. 825–836) describes an experiment that studies the relationship of the Mooney viscosity of rubber to several variables, including silica filler (parts per hundred) and oil filler (parts per hundred). Some of the data from this experiment are shown here, where

$$x_1 = \frac{silica - 60}{15} \qquad x_2 = \frac{oil - 21}{1.5}$$

Coded Levels		
x_1	x_2	y
−1	−1	13.71
1	−1	14.15
−1	1	12.87
1	1	13.53
−1.4	0	12.99
1.4	0	13.89
0	−1.4	14.16
0	1.4	12.90
0	0	13.75
0	0	13.66
0	0	13.86
0	0	13.63
0	0	13.74

(a) What type of experimental design has been used? Is it rotatable?

   The design is a CCD with $k = 2$ and $\alpha = 1.4$. The design is rotatable.

(b) Fit a quadratic model to these data. What values of x_1 and x_2 will maximize the Mooney viscosity?

   Since the standard order is provided, one approach to solving this exercise is to create a two-factor response surface design in MINITAB, then enter the data. Select **Stat>DOE>Response Surface>Create Response Surface Design**. Leave the design type as a 2-factor, central composite design. Select "**Designs**", highlight the design with five center points (13 runs), and enter a custom alpha value of exactly 1.4 (the rotatable design is $\alpha = 1.41421$). The worksheet is in run order, to change to standard order (and ease data entry) select **Stat>DOE>Display Design** and choose standard order. To analyze the experiment, select **Stat>DOE>Response Surface>Analyze Response Surface Design**. Select "**Terms**" and verify that a full quadratic model (A, B, A^2, B^2, AB) is selected.

```
Response Surface Regression: y versus x1, x2
The analysis was done using coded units.
Estimated Regression Coefficients for y
Term        Coef    SE Coef       T       P
Constant  13.7273   0.04309  318.580   0.000
x1         0.2980   0.03424    8.703   0.000
x2        -0.4071   0.03424  -11.889   0.000
x1*x1     -0.1249   0.03706   -3.371   0.012
x2*x2     -0.0790   0.03706   -2.132   0.070
x1*x2      0.0550   0.04818    1.142   0.291
...
Analysis of Variance for y
Source          DF   Seq SS   Adj SS   Adj MS      F       P
Regression       5  2.16128  2.16128  0.43226  46.56   0.000
  Linear         2  2.01563  2.01563  1.00781 108.54   0.000
  Square         2  0.13355  0.13355  0.06678   7.19   0.020
  Interaction    1  0.01210  0.01210  0.01210   1.30   0.291
Residual Error   7  0.06499  0.06499  0.00928
  Lack-of-Fit    3  0.03271  0.03271  0.01090   1.35   0.377
  Pure Error     4  0.03228  0.03228  0.00807
Total           12  2.22628
...
```

```
Estimated Regression Coefficients for y using data in uncoded units
Term        Coef
Constant  13.7273
x1         0.2980
x2        -0.4071
x1*x1     -0.1249
x2*x2     -0.0790
x1*x2      0.0550
```

Values of x_1 and x_2 maximizing the Mooney viscosity can be found from visual examination of the contour and surface plots, or using MINITAB's Response Optimizer.

Stat>DOE>Response Surface>Contour/Surface Plots

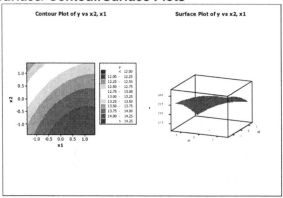

Stat>DOE>Response Surface>Response Optimizer
In Setup, let Goal = maximize, Lower = 10, Target = 20, and Weight = 7.

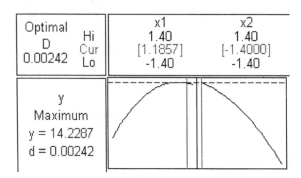

From the plots and the optimizer, setting x_1 in a range from 0 to +1.4 and setting x_2 between −1 and −1.4 will maximize viscosity.

13-7. An article by J. J. Pignatiello, Jr. and J. S. Ramberg in the Journal of Quality Technology (Vol. 17, 1985, pp. 198-206) describes the use of a replicated fractional factorial to investigate the effect of five factors on the free height of leaf springs used in an automotive application. The factors are A = furnace temperature, B = heating time, C = transfer time, D = hold down time, and E = quench oil temperature. The data are shown here.

A	B	C	D	E			
–	–	–	–	–	7.78,	7.78,	7.81
+	–	–	+	–	8.15,	8.18,	7.88
–	+	–	+	–	7.50,	7.56,	7.50
+	+	–	–	–	7.59,	7.56,	7.75
–	–	+	+	–	7.54,	8.00,	7.88
+	–	+	–	–	7.69,	8.09,	8.06
–	+	+	–	–	7.56,	7.52,	7.44
+	+	+	+	–	7.56,	7.81,	7.69
–	–	–	–	+	7.50,	7.25,	7.12
+	–	–	+	+	7.88,	7.88,	7.44
–	+	–	+	+	7.50,	7.56,	7.50
+	+	–	–	+	7.63,	7.75,	7.56
–	–	+	+	+	7.32,	7.44,	7.44
+	–	+	–	+	7.56,	7.69,	7.62
–	+	+	–	+	7.18,	7.18,	7.25
+	+	+	+	+	7.81,	7.50,	7.59

(a) Write out the alias structure for this design. What is the resolution of this design?

Enter the factor levels and response data into a MINITAB worksheet, and then define the experiment using **Stat>DOE>Factorial>Define Custom Factorial Design**. The defining relation for this half-fraction design is I = ABCD (from examination of the plus and minus signs).

A+BCD	AB+CD	CE+ABDE
B+ACD	AC+BD	DE+ABCE
C+ABD	AD+BC	ABE+CDE
D+ABC	AE+BCDE	ACE+BDE
E	BE+ACDE	ADE+BCE

This is a resolution IV design. All main effects are clear of 2-factor interactions, but some 2-factor interactions are aliased with each other.

Stat>DOE>Factorial>Analyze Factorial Design

```
Factorial Fit: Mean versus A, B, C, D, E
...
Alias Structure
I + A*B*C*D
A + B*C*D
B + A*C*D
C + A*B*D
D + A*B*C
E + A*B*C*D*E
A*B + C*D
A*C + B*D
A*D + B*C
A*E + B*C*D*E
B*E + A*C*D*E
C*E + A*B*D*E
D*E + A*B*C*E
```

13-7 continued

(b) Analyze the data. What factors influence mean free height?
The full model for mean:

Stat>DOE>Factorial>Analyze Factorial Design

Factorial Fit: Height versus A, B, C, D, E

Estimated Effects and Coefficients for Height (coded units)

Term	Effect	Coef	SE Coef	T	P
Constant		7.6256	0.02021	377.41	0.000
A	0.2421	0.1210	0.02021	5.99	0.000
B	-0.1638	-0.0819	0.02021	-4.05	0.000
C	-0.0496	-0.0248	0.02021	-1.23	0.229
D	0.0912	0.0456	0.02021	2.26	0.031
E	-0.2387	-0.1194	0.02021	-5.91	0.000
A*B	-0.0296	-0.0148	0.02021	-0.73	0.469
A*C	0.0012	0.0006	0.02021	0.03	0.976
A*D	-0.0229	-0.0115	0.02021	-0.57	0.575
A*E	0.0637	0.0319	0.02021	1.58	0.124
B*E	0.1529	0.0765	0.02021	3.78	0.001
C*E	-0.0329	-0.0165	0.02021	-0.81	0.421
D*E	0.0396	0.0198	0.02021	0.98	0.335
A*B*E	0.0021	0.0010	0.02021	0.05	0.959
A*C*E	0.0196	0.0098	0.02021	0.48	0.631
A*D*E	-0.0596	-0.0298	0.02021	-1.47	0.150

...

Analysis of Variance for Height (coded units)

Source	DF	Seq SS	Adj SS	Adj MS	F	P
Main Effects	5	1.83846	1.83846	0.36769	18.76	0.000
2-Way Interactions	7	0.37800	0.37800	0.05400	2.76	0.023
3-Way Interactions	3	0.04726	0.04726	0.01575	0.80	0.501
Residual Error	32	0.62707	0.62707	0.01960		
Pure Error	32	0.62707	0.62707	0.01960		
Total	47	2.89078				

The reduced model for mean:

Factorial Fit: Height versus A, B, D, E

Estimated Effects and Coefficients for Height (coded units)

Term	Effect	Coef	SE Coef	T	P
Constant		7.6256	0.01994	382.51	0.000
A	0.2421	0.1210	0.01994	6.07	0.000
B	-0.1638	-0.0819	0.01994	-4.11	0.000
D	0.0913	0.0456	0.01994	2.29	0.027
E	-0.2387	-0.1194	0.01994	-5.99	0.000
B*E	0.1529	0.0765	0.01994	3.84	0.000

...

Analysis of Variance for Height (coded units)

Source	DF	Seq SS	Adj SS	Adj MS	F	P
Main Effects	4	1.8090	1.8090	0.45224	23.71	0.000
2-Way Interactions	1	0.2806	0.2806	0.28060	14.71	0.000
Residual Error	42	0.8012	0.8012	0.01908		
Lack of Fit	10	0.1742	0.1742	0.01742	0.89	0.554
Pure Error	32	0.6271	0.6271	0.01960		
Total	47	2.8908				

13-7 continued

(c) Calculate the range and standard deviation of free height for each run. Is there any indication that any of these factors affects variability in free height?

The full model for range:

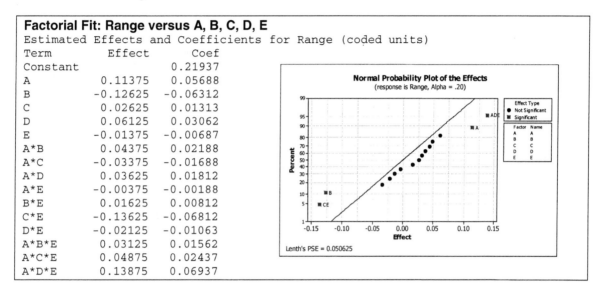

Factorial Fit: Range versus A, B, C, D, E

Estimated Effects and Coefficients for Range (coded units)

Term	Effect	Coef
Constant		0.21937
A	0.11375	0.05688
B	-0.12625	-0.06312
C	0.02625	0.01313
D	0.06125	0.03062
E	-0.01375	-0.00687
A*B	0.04375	0.02188
A*C	-0.03375	-0.01688
A*D	0.03625	0.01812
A*E	-0.00375	-0.00188
B*E	0.01625	0.00812
C*E	-0.13625	-0.06812
D*E	-0.02125	-0.01063
A*B*E	0.03125	0.01562
A*C*E	0.04875	0.02437
A*D*E	0.13875	0.06937

The reduced model for range:

Factorial Fit: Range versus A, B, C, D, E

Estimated Effects and Coefficients for Range (coded units)

Term	Effect	Coef	SE Coef	T	P
Constant		0.21937	0.01625	13.50	0.000
A	0.11375	0.05688	0.01625	3.50	0.008
B	-0.12625	-0.06312	0.01625	-3.88	0.005
C	0.02625	0.01313	0.01625	0.81	0.443
D	0.06125	0.03062	0.01625	1.88	0.096
E	-0.01375	-0.00687	0.01625	-0.42	0.683
C*E	-0.13625	-0.06812	0.01625	-4.19	0.003
A*D*E	0.13875	0.06937	0.01625	4.27	0.003

...

Analysis of Variance for Range (coded units)

Source	DF	Seq SS	Adj SS	Adj MS	F	P
Main Effects	5	0.13403	0.13403	0.026806	6.34	0.011
2-Way Interactions	1	0.07426	0.07426	0.074256	17.58	0.003
3-Way Interactions	1	0.07701	0.07701	0.077006	18.23	0.003
Residual Error	8	0.03380	0.03380	0.004225		
Total	15	0.31909				

...

13-7 continued

The full model for standard deviation:

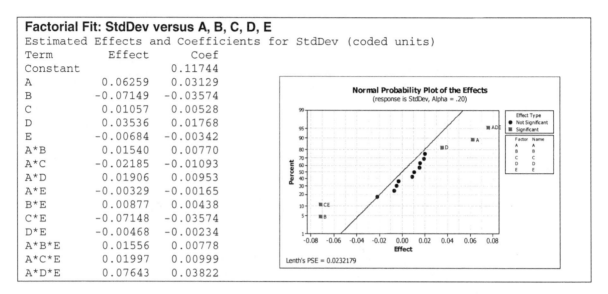

Factorial Fit: StdDev versus A, B, C, D, E		
Estimated Effects and Coefficients for StdDev (coded units)		
Term	Effect	Coef
Constant		0.11744
A	0.06259	0.03129
B	-0.07149	-0.03574
C	0.01057	0.00528
D	0.03536	0.01768
E	-0.00684	-0.00342
A*B	0.01540	0.00770
A*C	-0.02185	-0.01093
A*D	0.01906	0.00953
A*E	-0.00329	-0.00165
B*E	0.00877	0.00438
C*E	-0.07148	-0.03574
D*E	-0.00468	-0.00234
A*B*E	0.01556	0.00778
A*C*E	0.01997	0.00999
A*D*E	0.07643	0.03822

The reduced model for standard deviation:

Factorial Fit: StdDev versus A, B, C, D, E

Estimated Effects and Coefficients for StdDev (coded units)

Term	Effect	Coef	SE Coef	T	P
Constant		0.11744	0.007559	15.54	0.000
A	0.06259	0.03129	0.007559	4.14	0.003
B	-0.07149	-0.03574	0.007559	-4.73	0.001
C	0.01057	0.00528	0.007559	0.70	0.504
D	0.03536	0.01768	0.007559	2.34	0.047
E	-0.00684	-0.00342	0.007559	-0.45	0.663
C*E	-0.07148	-0.03574	0.007559	-4.73	0.001
A*D*E	0.07643	0.03822	0.007559	5.06	0.001

...

Analysis of Variance for StdDev (coded units)

Source	DF	Seq SS	Adj SS	Adj MS	F	P
Main Effects	5	0.041748	0.041748	0.0083496	9.13	0.004
2-Way Interactions	1	0.020438	0.020438	0.0204385	22.36	0.001
3-Way Interactions	1	0.023369	0.023369	0.0233690	25.56	0.001
Residual Error	8	0.007314	0.007314	0.0009142		
Total	15	0.092869				

For both models of variability, interactions CE (transfer time × quench oil temperature) and ADE=BCE, along with factors B (heating time) and A (furnace temperature) are significant. Factors C and E are included to keep the models hierarchical.

13-7 continued

(d) Analyze the residuals from this experiment and comment on your findings.

For mean height:

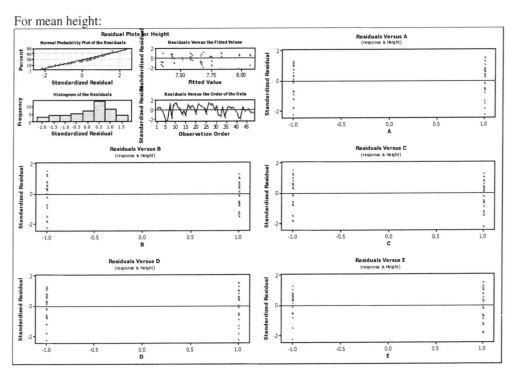

For range:

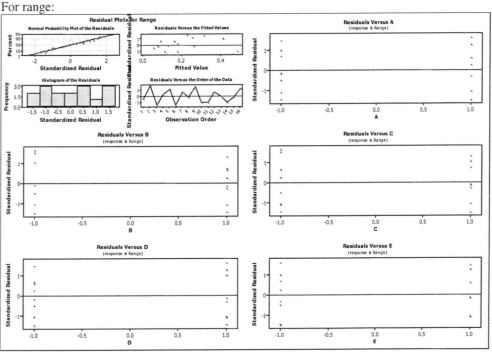

13-7 continued

For standard deviation:

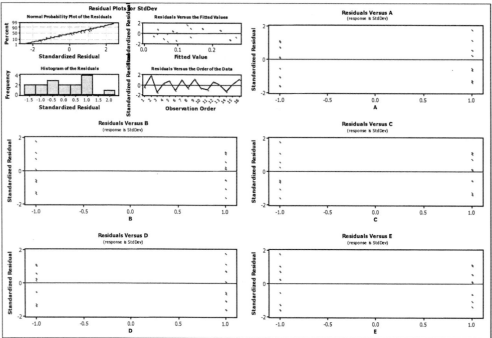

Mean Height

Plot of residuals versus predicted indicates constant variance assumption is reasonable. Normal probability plot of residuals support normality assumption. Plots of residuals versus each factor shows that variance is less at low level of factor E.

Range

Plot of residuals versus predicted shows that variance is approximately constant over range of predicted values. Residuals normal probability plot indicate normality assumption is reasonable Plots of residuals versus each factor indicate that the variance may be different at different levels of factor D.

Standard Deviation

Residuals versus predicted plot and residuals normal probability plot support constant variance and normality assumptions. Plots of residuals versus each factor indicate that the variance may be different at different levels of factor D.

(e) Is this the best possible design for five factors in 16 runs? Specifically, can you find a fractional design for five factors in 16 runs with higher resolution than this one?

This is not the best 16-run design for five factors. A resolution V design can be generated with E = ± ABCD, then none of the 2-factor interactions will be aliased with each other.

13-9.   Consider the leaf spring experiment in Exercise 13-7. Rework this problem, assuming that factors A, B, and C are easy to control but factors D and E are hard to control.

Factors D and E are noise variables. Assume $\sigma_D^2 = \sigma_E^2 = 1$. Using equations (13-6) and (13-7), the mean and variance are:

Mean Free Height = $7.63 + 0.12A - 0.081B$
Variance of Free Height = $\sigma_D^2 (+0.046)^2 + \sigma_E^2 (-0.12 + 0.077B)^2 + \sigma^2$

Using $\hat{\sigma}^2 = MS_E = 0.02$:

Variance of Free Height = $(0.046)^2 + (-0.12 + 0.077B)^2 + 0.02$

For the current factor levels, Free Height Variance could be calculated in the MINITAB worksheet, and then contour plots in factors A, B, and D could be constructed using the Graph > Contour Plot functionality. These contour plots could be compared with a contour plot of Mean Free Height, and optimal settings could be identified from visual examination of both plots. This approach is fully described in the solution to Exercise 13-12.

The overlaid contour plot below (constructed in Design-Expert) shows one solution with mean Free Height $\cong 7.50$ and minimum standard deviation of Free Height to be: A = -0.42 and B = 0.99.

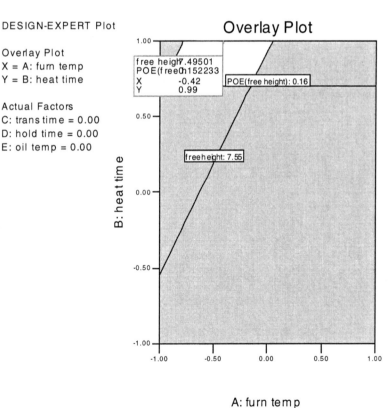

13-11. The following data were collected by a chemical engineer. The response y is filtration time, x_1 is temperature, and x_2 is pressure. Fit a second-order model.

x_1	x_2	y
-1	-1	54
-1	1	45
1	-1	32
1	1	47
-1.414	0	50
1.414	0	53
0	-1.414	47
0	1.414	51
0	0	41
0	0	39
0	0	44
0	0	42
0	0	40

Since the runs are listed in a patterned (but not standard) order, one approach to solving this exercise is to create a general full factorial design in MINITAB, and then enter the data.

Stat > DOE > Response Surface > Analyze Response Surface Design

Response Surface Regression: y versus x1, x2
```
The analysis was done using coded units.
Estimated Regression Coefficients for y
Term        Coef   SE Coef      T       P
Constant  41.200    2.100   19.616   0.000
x1        -1.970    1.660   -1.186   0.274
x2         1.457    1.660    0.878   0.409
x1*x1      3.712    1.781    2.085   0.076
x2*x2      2.463    1.781    1.383   0.209
x1*x2      6.000    2.348    2.555   0.038
...
Analysis of Variance for y
Source          DF   Seq SS   Adj SS    Adj MS      F      P
Regression       5   315.60   315.60    63.119   2.86  0.102
  Linear         2    48.02    48.02    24.011   1.09  0.388
  Square         2   123.58   123.58    61.788   2.80  0.128
  Interaction    1   144.00   144.00   144.000   6.53  0.038
Residual Error   7   154.40   154.40    22.058
  Lack-of-Fit    3   139.60   139.60    46.534  12.58  0.017
  Pure Error     4    14.80    14.80     3.700
Total           12   470.00
```

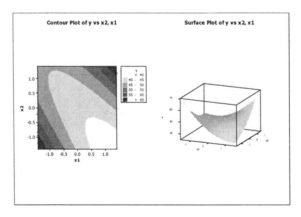

13-11 continued

(a) What operating conditions would you recommend if the objective is to minimize the filtration time?

Goal = Minimize, Target = 0, Upper = 55, Weight = 1, Importance = 1

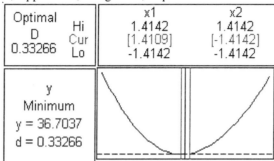

Optimal		x1	x2
	Hi	1.4142	1.4142
D	Cur	[1.4109]	[-1.4142]
0.33266	Lo	-1.4142	-1.4142

y	
Minimum	
y = 36.7037	
d = 0.33266	

Recommended operating conditions are temperature = +1.4109 and pressure = -1.4142, to achieve predicted filtration time of 36.7.

(b) What operating conditions would you recommend if the objective is to operate the process at a mean filtration rate very close to 46?

Goal = Target, Lower = 42, Target = 46, Upper = 50, Weight = 10, Importance = 1

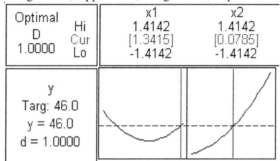

Optimal		x1	x2
	Hi	1.4142	1.4142
D	Cur	[1.3415]	[0.0785]
1.0000	Lo	-1.4142	-1.4142

y	
Targ: 46.0	
y = 46.0	
d = 1.0000	

Recommended operating conditions are temperature = +1.3415 and pressure = -0.0785, to achieve predicted filtration time of 46.0.

CHAPTER 14

Lot-by-Lot Acceptance Sampling for Attributes

Learning Objectives

After completing this chapter you should be able to:
1. Understand the role of acceptance sampling in modern quality control systems
2. Understand the advantages and disadvantages of sampling
3. Understand the difference between attributes and variables sampling plans, and the major types of acceptance-sampling procedures
4. Know how single-, double-, and sequential-sampling plans are used
5. Understand the importance of random sampling
6. Know how to determine the OC curve for a single-sampling plan for attributes
7. Understand the effects of the sampling plan parameters on sampling plan performance
8. Know how to design single-sampling, double-sampling, and sequential sampling plans for attributes
9. Know how rectifying inspection is used
10. Understand the structure and use of MIL STD 105E and its civilian counterpart plans
11. Understand the structure and use of the Dodge-Romig system of sampling plans

Important Terms and Concepts

100% inspection

Acceptance-sampling plan

AOQL plans

Average outgoing quality

Average sample number curve

Dodge-Romig sampling plans

Ideal OC curve

Lot sentencing

LTPD plans

Multiple-sampling plan

Operating-characteristic (OC) curve

Rectifying inspection

Sequential-sampling plan

Switching rules in MIL STD 105E

Variables data

Acceptable quality level (AQL)

ANSI/ASQC Z1.4, ISO 2859

Attributes data

Average outgoing quality limit

Average total inspection

Double-sampling plan

Lot disposition actions

Lot tolerance percent defective (LTPD)

MIL STD 105E

Normal, tightened, and reduced inspection

Random sampling

Sample size code letters

Single-sampling plan

Type-A and Type-B OC Curves

Exercises

Note: Many of the exercises in this chapter are easily solved with spreadsheet application software. The BINOMDIST, HYPGEOMDIST, and graphing functions in Microsoft® Excel were used for these solutions.

14-1.   Draw the type-B OC curve for the single-sampling plan $n = 50$, $c = 1$.

Binomial distribution and

$$P_a = P\{d \le c\} = \sum_{d=0}^{c} \frac{n!}{d!(n-d)!} p^d (1-p)^{n-d}$$ (14-2)

Excel formulas:

	A	B	C	D
1	n =	50		
2	c =	1		
3	d =	0	1	
4	p	f(d=0)	f(d=1)	Pr{d<=c}
5	0.001	=FACT($B$1)/(FACT(B$3)*FACT($B$1-B$3))*$A5^B$3*(1-$A5)^($B$1-B$3)	=FACT($B$1)/(FACT(C$3)*FACT($B$1-C$3))*$A5^C$3*(1-$A5)^($B$1-C$3)	=+B5+C5
6	0.002	=FACT($B$1)/(FACT(B$3)*FACT($B$1-B$3))*$A6^B$3*(1-$A6)^($B$1-B$3)	=FACT($B$1)/(FACT(C$3)*FACT($B$1-C$3))*$A6^C$3*(1-$A6)^($B$1-C$3)	=+B6+C6
7	0.003	=FACT($B$1)/(FACT(B$3)*FACT($B$1-B$3))*$A7^B$3*(1-$A7)^($B$1-B$3)	=FACT($B$1)/(FACT(C$3)*FACT($B$1-C$3))*$A7^C$3*(1-$A7)^($B$1-C$3)	=+B7+C7
8	0.004	=FACT($B$1)/(FACT(B$3)*FACT($B$1-B$3))*$A8^B$3*(1-$A8)^($B$1-B$3)	=FACT($B$1)/(FACT(C$3)*FACT($B$1-C$3))*$A8^C$3*(1-$A8)^($B$1-C$3)	=+B8+C8
9	0.005	=FACT($B$1)/(FACT(B$3)*FACT($B$1-B$3))*$A9^B$3*(1-$A9)^($B$1-B$3)	=FACT($B$1)/(FACT(C$3)*FACT($B$1-C$3))*$A9^C$3*(1-$A9)^($B$1-C$3)	=+B9+C9
10	0.006	=FACT($B$1)/(FACT(B$3)*FACT($B$1-B$3))*$A10^B$3*(1-$A10)^($B$1-B$3)	=FACT($B$1)/(FACT(C$3)*FACT($B$1-C$3))*$A10^C$3*(1-$A10)^($B$1-C$3)	=+B10+C10
11	...			

Excel results:

	A	B	C	D
1	n =	50		
2	c =	1		
3	d =	0	1	
4	p	f(d=0)	f(d=1)	Pr{d<=c}
5	0.001	0.95121	0.04761	0.99881
6	0.002	0.90475	0.09066	0.99540
7	0.003	0.86051	0.12947	0.98998
8	0.004	0.81840	0.16434	0.98274
9	0.005	0.77831	0.19556	0.97387
10	0.006	0.74015	0.22339	0.96353
11	...			

Excel graph:

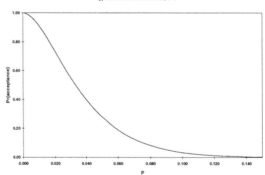

14-3.   Suppose that a product is shipped in lots of size $N = 5000$. The receiving inspection procedure used is single sampling with $n = 50$ and $c = 1$.

Hypergeometric distribution and

$$P_a = P\{d \le c\} = \sum_{d=0}^{c} \frac{n!}{d!(n-d)!} p^d (1-p)^{n-d} \qquad (14\text{-}2)$$

(a) Draw the type-A OC curve for the plan.

Excel formulas:

	A	E	F	G	H
1	N =	5000			
2	n =	50			
3	c =	1			
4	d =	0	1		
5					
6	p	D	f(d=0)	f(d=1)	Pr{d<=1}
7	0.001	=INT(D7)	=HYPGEOMDIST(E$4,$E$2,$E7,$E$1)	=HYPGEOMDIST(F$4,$E$2,$E7,$E$1)	=F7+G7
8	0.002	=INT(D8)	=HYPGEOMDIST(E$4,$E$2,$E8,$E$1)	=HYPGEOMDIST(F$4,$E$2,$E8,$E$1)	=F8+G8
9	0.003	=INT(D9)	=HYPGEOMDIST(E$4,$E$2,$E9,$E$1)	=HYPGEOMDIST(F$4,$E$2,$E9,$E$1)	=F9+G9
10	0.004	=INT(D10)	=HYPGEOMDIST(E$4,$E$2,$E10,$E$1)	=HYPGEOMDIST(F$4,$E$2,$E10,$E$1)	=F10+G10
11	0.005	=INT(D11)	=HYPGEOMDIST(E$4,$E$2,$E11,$E$1)	=HYPGEOMDIST(F$4,$E$2,$E11,$E$1)	=F11+G11
12	0.006	=INT(D12)	=HYPGEOMDIST(E$4,$E$2,$E12,$E$1)	=HYPGEOMDIST(F$4,$E$2,$E12,$E$1)	=F12+G12
13	0.007	=INT(D13)	=HYPGEOMDIST(E$4,$E$2,$E13,$E$1)	=HYPGEOMDIST(F$4,$E$2,$E13,$E$1)	=F13+G13
14	0.0075	=INT(D14)	=HYPGEOMDIST(E$4,$E$2,$E14,$E$1)	=HYPGEOMDIST(F$4,$E$2,$E14,$E$1)	=F14+G14
15	0.008	=INT(D15)	=HYPGEOMDIST(E$4,$E$2,$E15,$E$1)	=HYPGEOMDIST(F$4,$E$2,$E15,$E$1)	=F15+G15
16	0.009	=INT(D16)	=HYPGEOMDIST(E$4,$E$2,$E16,$E$1)	=HYPGEOMDIST(F$4,$E$2,$E16,$E$1)	=F16+G16
17	...				

Excel results:

	A	B
5		binomial - Type B
6	p	Pr{d<=1}
7	0.001	0.99881
8	0.002	0.99540
9	0.003	0.98998
10	0.004	0.98274
11	0.005	0.97387
12	0.006	0.96353
13	0.007	0.95190
14	0.008	0.94563
15	0.008	0.93910
16	0.009	0.92528
17	0.010	0.91056
18	...	

$P_a (d = 35) = 0.9521$, or $\alpha \cong 0.05$ and $P_a (d = 375) = 0.10133$, or $\beta \cong 0.10$

Excel graph:

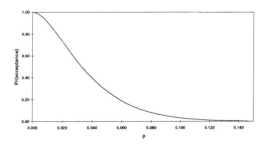

Type-A OC Curve for N=5000, n=50, c=1

14-3 continued

(b) Draw the type-B OC curve for this plan and compare it to the type-A OC curve found in part (a).

Excel formulas:

	A	B	C	D	E
1	N =	5000			
2	n =	50			
3	c =	1			
4	d =	0	1		
5		binomial - Type B				
6	p	Pr{d<=1}		N*p	D
7	0.001	=BINOMDIST($B$3,$B$2,A7,TRUE)		=$B$1*A7	=INT(D7)
8	0.002	=BINOMDIST($B$3,$B$2,A8,TRUE)		=$B$1*A8	=INT(D8)
9	0.003	=BINOMDIST($B$3,$B$2,A9,TRUE)		=$B$1*A9	=INT(D9)
10	0.004	=BINOMDIST($B$3,$B$2,A10,TRUE)		=$B$1*A10	=INT(D10)
11	0.005	=BINOMDIST($B$3,$B$2,A11,TRUE)		=$B$1*A11	=INT(D11)
12	0.006	=BINOMDIST($B$3,$B$2,A12,TRUE)		=$B$1*A12	=INT(D12)
13	0.007	=BINOMDIST($B$3,$B$2,A13,TRUE)		=$B$1*A13	=INT(D13)
14	0.0075	=BINOMDIST($B$3,$B$2,A14,TRUE)		=$B$1*A14	=INT(D14)
15	0.008	=BINOMDIST($B$3,$B$2,A15,TRUE)		=$B$1*A15	=INT(D15)
16	0.009	=BINOMDIST($B$3,$B$2,A16,TRUE)		=$B$1*A16	=INT(D16)

Excel results:

	A	D	E	F	G	H
5			hypergeometric - Type A			
6	p	N*p	D	f(d=0)	f(d=1)	Pr{d<=1}
7	0.001	5	5	0.95097	0.04807	0.99904
8	0.002	10	10	0.90430	0.09151	0.99581
9	0.003	15	15	0.85988	0.13065	0.99053
10	0.004	20	20	0.81759	0.16581	0.98340
11	0.005	25	25	0.77735	0.19726	0.97461
12	0.006	30	30	0.73905	0.22527	0.96432
13	0.007	35	35	0.70260	0.25011	0.95271
14	0.008	37.5	37	0.68852	0.25921	0.94773
15	0.008	40	40	0.66791	0.27201	0.93992
16	0.009	45	45	0.63491	0.29118	0.92609
17	0.010	50	50	0.60350	0.30785	0.91135
18	...					

$P_a (p = 0.007) = 0.9521$, or $\alpha \cong 0.05$ and $P_a (p = 0.075) = 0.10133$, or $\beta \cong 0.10$

Excel graph:

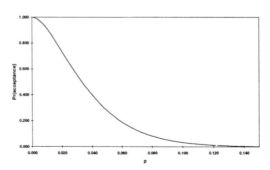

Type-B OC Curve for N=5000, n=50, c=1

(c) Which curve is appropriate for this situation?

Based on values for α and β, the difference between the two curves is small; either is appropriate.

14-5. Find a single-sampling plan for which $p_1 = 0.05$, $\alpha = 0.05$, $p_2 = 0.15$, and $\beta = 0.10$.

$p_1 = 0.05; 1 - \alpha = 1 - 0.05 = 0.95; p_2 = 0.15; \beta = 0.10$

From the binomial nomograph (Figure 14-9), the sampling plan is $n = 80$ and $c = 7$.

14-7. A company uses the following acceptance-sampling procedure. A sample equal to 10% of the lot is taken. If 2% or less of the items in the sample are defective, the lot is accepted; otherwise, it is rejected. If submitted lots vary in size from 5000 to 10,000 units, what can you say about the protection by this plan? If 0.05 is the desired LTPD, does this scheme offer reasonable protection to the consumer?

Excel formulas:

	A	B	C	D	E	F
1	LTPD =	0.05				
2						
3		N1 =	5000		N2 =	10000
4		n1 =	=0.1*C3		n1 =	=0.1*F3
5		pmax =	0.02		pmax =	0.02
6		cmax =	=C5*C4		cmax =	=F5*F4
7		binomial			binomial	
8	p	Pr{d<=10}	Pr{reject}		Pr{d<=20}	Pr{reject}
9	0.001	=BINOMDIST($C$6,$C$4,A9,TRUE)	=1-B9		=BINOMDIST($F$6,$F$4,A9,TRUE)	=1-E9
10	0.002	=BINOMDIST($C$6,$C$4,A10,TRUE)	=1-B10		=BINOMDIST($F$6,$F$4,A10,TRUE)	=1-E10
11	0.003	=BINOMDIST($C$6,$C$4,A11,TRUE)	=1-B11		=BINOMDIST($F$6,$F$4,A11,TRUE)	=1-E11
12	0.004	=BINOMDIST($C$6,$C$4,A12,TRUE)	=1-B12		=BINOMDIST($F$6,$F$4,A12,TRUE)	=1-E12
13	0.005	=BINOMDIST($C$6,$C$4,A13,TRUE)	=1-B13		=BINOMDIST($F$6,$F$4,A13,TRUE)	=1-E13
14	0.006	=BINOMDIST($C$6,$C$4,A14,TRUE)	=1-B14		=BINOMDIST($F$6,$F$4,A14,TRUE)	=1-E14
15	0.007	=BINOMDIST($C$6,$C$4,A15,TRUE)	=1-B15		=BINOMDIST($F$6,$F$4,A15,TRUE)	=1-E15
16	0.0075	=BINOMDIST($C$6,$C$4,A16,TRUE)	=1-B16		=BINOMDIST($F$6,$F$4,A16,TRUE)	=1-E16
17	0.008	=BINOMDIST($C$6,$C$4,A17,TRUE)	=1-B17		=BINOMDIST($F$6,$F$4,A17,TRUE)	=1-E17
18	0.009	=BINOMDIST($C$6,$C$4,A18,TRUE)	=1-B18		=BINOMDIST($F$6,$F$4,A18,TRUE)	=1-E18
19	0.01	=BINOMDIST($C$6,$C$4,A19,TRUE)	=1-B19		=BINOMDIST($F$6,$F$4,A19,TRUE)	=1-E19
20	...					

Excel results:

	A	B	C	D	E	F	G	H
8	p	Pr{d<=10}	Pr{reject}		Pr{d<=20}	Pr{reject}		difference
9	0.0010	1.00000	0.0000		1.00000	0.0000		0.00000
10	0.0020	1.00000	0.0000		1.00000	0.0000		0.00000
11	0.0030	1.00000	0.0000		1.00000	0.0000		0.00000
12	0.0040	0.99999	0.0000		1.00000	0.0000		-0.00001
13	0.0050	0.99994	0.0001		1.00000	0.0000		-0.00006
14	0.0060	0.99972	0.0003		1.00000	0.0000		-0.00027
15	0.0070	0.99903	0.0010		0.99999	0.0000		-0.00095
16	0.0075	0.99834	0.0017		0.99996	0.0000		-0.00163
17	0.0080	0.99729	0.0027		0.99991	0.0001		-0.00263
18	0.0090	0.99359	0.0064		0.99959	0.0004		-0.00600
19	0.0100	0.98676	0.0132		0.99850	0.0015		-0.01175
20	0.0110	0.97545	0.0245		0.99556	0.0044		-0.02010
21	0.0120	0.95837	0.0416		0.98886	0.0111		-0.03049
22	0.0130	0.93444	0.0656		0.97579	0.0242		-0.04135
23	0.0140	0.90298	0.0970		0.95330	0.0467		-0.05031
24	0.0150	0.86386	0.1361		0.91861	0.0814		-0.05474
25	0.0200	0.58304	0.4170		0.55910	0.4409		0.02395
26	0.0250	0.29404	0.7060		0.18221	0.8178		0.11183
27	0.0300	0.11479	0.8852		0.03328	0.9667		0.08151
28	0.0350	0.03631	0.9637		0.00380	0.9962		0.03251
29	0.0400	0.00967	0.9903		0.00030	0.9997		0.00938
30	0.0450	0.00224	0.9978		0.00002	1.0000		0.00222
31	0.0500	0.00046	0.9995		0.00000	1.0000		0.00046
32	...							

Different sample sizes offer different levels of protection. For $N = 5,000$, $P_a(p = 0.025) = 0.294$; while for $N = 10,000$, $P_a(p = 0.025) = 0.182$. Also, the consumer is protected from a LTPD = 0.05 by $P_a(N = 5,000) = 0.00046$ and $P_a(N = 10,000) = 0.00000$, but pays for the high probability of rejecting acceptable lots like those with $p = 0.025$.

14-9. Consider the single-sampling plan found in Exercise 14-4. Suppose that lots of $N = 2000$ are submitted. Draw the ATI curve for this plan. Draw the AOQ curve and find the AOQL.

$$n = 35; \ c = 1; \ N = 2,000$$

$$
\begin{aligned}
\text{ATI} &= n + (1 - P_a)(N - n) \\
&= 35 + (1 - P_a)(2000 - 35) \\
&= 2000 - 1965 P_a
\end{aligned}
$$

$$
\begin{aligned}
\text{AOQ} &= \frac{P_a \, p(N - n)}{N} \\
&= (1965/2000) P_a \, p
\end{aligned}
$$

$$\text{AOQL} = 0.0234$$

Excel formulas:

	A	B	C	D	E	F	G
2	n =	35					
3	c =	1					
4		binomial					
5	p	Pa=Pr{d<=1}		ATI		AOQ	
6	0.001	=BINOMDIST($B$3,$B$2,A6,TRUE)		=$B$1-($B$1-$B$2)*B6		=B6*A6*($B$1-$B$2)/$B$1	
7	0.002	=BINOMDIST($B$3,$B$2,A7,TRUE)		=$B$1-($B$1-$B$2)*B7		=B7*A7*($B$1-$B$2)/$B$1	
8	0.003	=BINOMDIST($B$3,$B$2,A8,TRUE)		=$B$1-($B$1-$B$2)*B8		=B8*A8*($B$1-$B$2)/$B$1	
9	0.004	=BINOMDIST($B$3,$B$2,A9,TRUE)		=$B$1-($B$1-$B$2)*B9		=B9*A9*($B$1-$B$2)/$B$1	
10	0.005	=BINOMDIST($B$3,$B$2,A10,TRUE)		=$B$1-($B$1-$B$2)*B10		=B10*A10*($B$1-$B$2)/$B$1	
11	0.006	=BINOMDIST($B$3,$B$2,A11,TRUE)		=$B$1-($B$1-$B$2)*B11		=B11*A11*($B$1-$B$2)/$B$1	
12	0.007	=BINOMDIST($B$3,$B$2,A12,TRUE)		=$B$1-($B$1-$B$2)*B12		=B12*A12*($B$1-$B$2)/$B$1	
13	0.0075	=BINOMDIST($B$3,$B$2,A13,TRUE)		=$B$1-($B$1-$B$2)*B13		=B13*A13*($B$1-$B$2)/$B$1	
14	0.008	=BINOMDIST($B$3,$B$2,A14,TRUE)		=$B$1-($B$1-$B$2)*B14		=B14*A14*($B$1-$B$2)/$B$1	
15	0.009	=BINOMDIST($B$3,$B$2,A15,TRUE)		=$B$1-($B$1-$B$2)*B15		=B15*A15*($B$1-$B$2)/$B$1	
16	0.01	=BINOMDIST($B$3,$B$2,A16,TRUE)		=$B$1-($B$1-$B$2)*B16		=B16*A16*($B$1-$B$2)/$B$1	
17	0.011	=BINOMDIST($B$3,$B$2,A17,TRUE)		=$B$1-($B$1-$B$2)*B17		=B17*A17*($B$1-$B$2)/$B$1	
18	0.012	=BINOMDIST($B$3,$B$2,A18,TRUE)		=$B$1-($B$1-$B$2)*B18		=B18*A18*($B$1-$B$2)/$B$1	
19	0.013	=BINOMDIST($B$3,$B$2,A19,TRUE)		=$B$1-($B$1-$B$2)*B19		=B19*A19*($B$1-$B$2)/$B$1	
20	0.014	=BINOMDIST($B$3,$B$2,A20,TRUE)		=$B$1-($B$1-$B$2)*B20		=B20*A20*($B$1-$B$2)/$B$1	
21	0.015	=BINOMDIST($B$3,$B$2,A21,TRUE)		=$B$1-($B$1-$B$2)*B21		=B21*A21*($B$1-$B$2)/$B$1	
22	0.02	=BINOMDIST($B$3,$B$2,A22,TRUE)		=$B$1-($B$1-$B$2)*B22		=B22*A22*($B$1-$B$2)/$B$1	
23	0.025	=BINOMDIST($B$3,$B$2,A23,TRUE)		=$B$1-($B$1-$B$2)*B23		=B23*A23*($B$1-$B$2)/$B$1	
24	0.03	=BINOMDIST($B$3,$B$2,A24,TRUE)		=$B$1-($B$1-$B$2)*B24		=B24*A24*($B$1-$B$2)/$B$1	
25	0.035	=BINOMDIST($B$3,$B$2,A25,TRUE)		=$B$1-($B$1-$B$2)*B25		=B25*A25*($B$1-$B$2)/$B$1	
26	0.04	=BINOMDIST($B$3,$B$2,A26,TRUE)		=$B$1-($B$1-$B$2)*B26		=B26*A26*($B$1-$B$2)/$B$1	
27	0.045	=BINOMDIST($B$3,$B$2,A27,TRUE)		=$B$1-($B$1-$B$2)*B27		=B27*A27*($B$1-$B$2)/$B$1	AOQL
28	...						